Bob Weil

D1222649

Spectral Analysis of Time-Series Data

METHODOLOGY IN THE SOCIAL SCIENCES
David A. Kenny, *Series Editor*

PRINCIPLES AND PRACTICE OF
STRUCTURAL EQUATION MODELING
Rex B. Kline

SPECTRAL ANALYSIS
OF TIME-SERIES DATA
Rebecca M. Warner

Spectral Analysis of Time-Series Data

REBECCA M. WARNER
University of New Hampshire

Series Editor's Note by David A. Kenny

THE GUILFORD PRESS
New York   London

© 1998 The Guilford Press
A Division of Guilford Publications, Inc.
72 Spring Street, New York, NY 10012
http://www.guilford.com

Printed in the United States of America

This book is printed on acid-free paper.

Last digit is print number: 9 8 7 6 5 4 3 2 1

Library of Congress Cataloging-in-Publication Data

Warner, Rebecca M.
    Spectral analysis of time-series data / Rebecca M. Warner.
        p.   cm. — (Methodology in the social sciences)
    Includes bibliographical references (p.   ) and index.
    ISBN 1-57230-338-7
    1. Time-series analysis.   2. Spectral theory (Mathematics)
    I. Title.   II. Series.
    QA280.W37   1998
    519.5'5—dc21   97-44005
                                                    CIP

Series Editor's Note

I am proud to introduce *Spectral Analysis of Time-Series Data* by Rebecca M. Warner for The Guilford Press series "Methodology in the Social Sciences." We have all heard the following expressions:

Turn, turn, turn.
What goes around comes around.
What goes up must come down.
Life is a circle.

All of them convey the idea that over-time phenomena are cyclical. However, few social or behavioral scientists know how to study cycles in data. At last, we have a book that speaks to us about how to estimate, test, and model cycles.

When presented with a time series, most of us often fit trends (e.g, linear, quadratic, cubic). We fail to realize that trends imply that the long-range forecast is a very extreme response; that is, trend models inevitably predict very extreme responses in the future. However, today, which is usually not all that extreme, was yesterday's future. Trends very often are insufficient in modeling over-time processes. Cycles often offer a much more reasonable way to understand variation over time than do trends.

The presentation of spectral analysis is often bogged down in incredibly complicated mathematics. It is filled with complex and even imaginary numbers. Although there is a beauty and an elegance in this mathematics, which Rebecca Warner appreciates, it is beyond the level of appreciation of most of us. This book develops some of the mathematics, but, more importantly, it tells the practicing researcher how to make sense out of the vast, and potentially bewildering, array of numbers from spectral analysis.

More and more, social and behavioral scientists are viewing behavior as determined by biological processes. The increasing influence is seen in cognition, behavioral medicine, personality, and relationship research, as well as several other areas. No doubt, advances in the mapping of human DNA will further increase the influence of biological models. A dominant mathematical technique employing biological models of over-time data is spectral analysis. Knowledge of this technique is required if we are to see a complete integration of the biological and the psychological.

One strong emphasis in *Spectral Analysis of Time-Series Data* is research design. How and how many observations are collected affects interpretation of the statistics. This book will become the "Campbell and Stanley" of cyclical analysis. That is, it carefully considers plausible rival hypotheses in the design and analysis of over-time research.

Among the strong features of the book are its many examples. The reader can reanalyze those data and reproduce the results presented in the book. Most of us learn from examples, and the book takes advantage of this fact.

I have known Rebecca Warner for over 20 years. Perhaps more than anyone else I know, she has shown a single-minded dedication to a focused set of problems. This book reflects the care and commitment of a scholar who has devoted her entire career to a significant and difficult problem. We are privileged to share in the wisdom that she has accumulated over the years.

DAVID A. KENNY
University of Connecticut, Storrs

Preface

The goal of this book is to provide interested readers with an understanding of several closely related methods of data analysis that describe cyclic patterns in time-series data. These methods are potentially useful for researchers or students who have many different kinds of time-series data: social indicator data (e.g., the number of divorces per year); systematically coded observational data (such as the level of affective involvement of each person in a mother–infant dyad); physiological data (such as measures of blood pressure); or measures of perceptual sensitivity or threshhold. Many (although not all) behavioral and physiological processes tend to be rhythmically organized; the cycles are often irregular and unstable over time, and they may not be easy to detect or describe. However, use of the methods described in this book provides some relatively simple ways to characterize any cyclic tendencies that are present (what proportion of variance in the time series is accounted for by the cycle? what is the length of the cycle? what is the amplitude of the cycle?).

Chapters 1, 2, and 3 outline issues in design of nonexperimental time-series studies, as well as preliminary examination of a single set of time-series data. Next, Chapters 4, 5, and 6 describe a set of closely related methods for the description of cycles in a single time series (harmonic analysis, periodogram analysis, and spectral analysis, respectively). For researchers who have just one time series, this may be sufficient.

However, many researchers have multiple time series. They may have time-series data on the same variable for many subjects (e.g., extensive time-series data on heart rate for each of 50 people). In this case, the researcher may want to evaluate whether the cycles in the time series are similar or different across subjects, and to assess whether any individual subject differences in the cycles are systematically related to outside variables such as subject gender or manipulated situational factors. Chapter 7 presents various ways of summarizing results across many univariate time-series analyses.

Alternatively, researchers may have time-series data on two or more variables for one subject (e.g., measures of heart rate, systolic blood pressure, and amount of talk for one person). The research questions then focus on relations between these variables over time; for instance, when the person begins to talk, does heart rate tend to increase soon afterward? If there are cycles in a physiological variable such as blood pressure, are these related to cycles in a behavioral variable, such as the amount of talking? Chapters 8, 9, and 10 outline methods for the analysis of bivariate time-series problems using lagged cross-correlations, simple time-series regression models, and cross-spectral analysis. Chapter 11 reviews some common sources of artifact and problems of interpretation that should be kept in mind when the researcher is doing either univariate or bivariate time-series analysis.

Finally, Chapter 12 describes some relevant theoretical issues: Why are some physiological and behavioral variables cyclic? What questions do researchers typically need to answer when they look for cycles in time-series data?

This book differs in several ways from most existing books on spectral analysis. First, the approach is much less mathematical. The focus here is on application and interpretation of spectral analysis and related analytic methods (not on derivations). Second, this book makes a case for treating the cyclic component of a time series as a potentially interesting pattern; some existing spectral analysis books focus mainly on removal of cycles from time series prior to doing other analyses. And, third, this book extensively discusses the problem of summarizing and comparing results of spectral analysis across time series for many different subjects, a problem that commonly arises in behavioral science research and that has received relatively little systematic attention in textbooks.

My hope is that readers who already have time-series data will find this a helpful guide to understanding their data. I also hope that this book will encourage researchers who have not yet collected time-series data to begin collecting time series in a broader variety of research situations. The recent availability of better-packaged programs that perform spectral analysis (and related analyses) will also do much to promote more widespread use of these methods. The SPSS for Windows TRENDS programs were used for all of the analyses presented as examples in this book, but many other packages offer good spectral analysis capabilities.

I would like to thank many people whose help made this book possible. I am indebted to the many statistics teachers whose courses challenged and inspired me and gave me the confidence to believe that I could contribute something of value. I am particularly grateful to my graduate school mentors Dr. David Kenny and Dr. Robert Rosenthal, without whom this book would not have been written. I owe great thanks also to many other teachers including the late Morris DeGroot and Dr. Richard Kronauer. I am thankful to the many students who have taken my statistics courses from 1981 to the present: a great deal of what I know about statistics has come from teaching and especially from trying to give good answers to student questions. Special thanks also to my parents (my first teachers). Finally, I appreciate the patience and moral support from my husband, Ed.

I also want to thank several organizations that supported the work for this book. I am particularly thankful to the University of New Hampshire for financial support through the UNH Faculty Fellowship Program and the UNH Summer Faculty Fellowship Program. Special thanks also to Victor Benassi, Chair of the Psychology Department at UNH, for small discretionary grants that paid for computer hardware and software and the occasional emergency computer repair. Some of the blood pressure and vocal activity data used in examples in this book were collected as part of a National Science Foundation–sponsored research grant (NSF Grant No. BNS-8819879).

REBECCA M. WARNER
University of New Hampshire

Contents

Research Questions for Time-Series and Spectral Analysis Studies

Introduction

Chapters 1 and 2 of this book deal in a relatively nontechnical manner with some questions about time-series research that need to be considered before a researcher can decide whether a time-series study is appropriate for the research situation at hand. Chapter 1 describes the types of data for which a time-series analytic approach is most appropriate. It outlines a strategy for data analysis, beginning with visual examination of a graph of the time-series data. The details for each step of this recommended data analysis strategy are provided in Chapters 3 through 6. Chapter 1 describes some of the patterns that a researcher can look for in time-series data, such as cycles.

Time-series novices will probably find it helpful to read the chapters of this book in order. Each chapter uses terminology and concepts from earlier chapters in a cumulative way. Novices may also find it helpful to enter the time-series data that are used in the examples into a time-series analysis program such as the SPSS for Windows TRENDS program, and duplicate the analyses that are presented in the text.

More experienced users may be able to skip early chapters (that deal with vocabulary and concepts that may already be familiar to them). Those who are interested specifically in interpretation and significance testing for univariate spectral analysis may be able to start at Chapter 6. Researchers who have already done extensive work with univariate time series but who are relatively unfamiliar with bivariate methods to assess relations between time series may want to begin with Chapter 8.

Time-Series Data Types

Many kinds of social and behavioral science research involve collection of time-series data. For instance, in studies of social indicators such as divorce, homicide, or suicide, investigators often obtain time-series data such as number of homicides per month for long periods of time from archival sources. In field observational studies, observers may code the frequency of specific behaviors at regular time intervals; for instance, the number of aggressive actions in a classroom may be coded once every 5 minutes. In longitudinal developmental studies, a psychologist may make repeated assessments of children once per year over many years to look at developmental trends. In laboratory studies, automated systems or human coders are often used to obtain behavioral or physiological time-series data, such as systolic blood pressure measured once per minute or level of self-disclosure rated by an observer once per minute.

Time-series data can be collected on many different kinds of behavior. Time-series social indicator data, such as rates of homicide or divorce, can be taken from archival records to monitor rates of behavior in large-scale social systems. Most of the examples discussed here focus on applications to individual persons or to small-scale social systems, such as face-to-face social interactions. Researchers who are interested in social interaction processes may collect objective data over time on expressive behaviors in adult–adult or infant–adult social interaction such as amount of vocal activity (Warner, 1988, 1991), gross motor body movement (Robertson, Dierker, Sorokin, & Rosen, 1982), or heart rate (Wade, Ellis, & Bohrer, 1973). Qualitative ratings may be obtained (from participants or from outside observers) over time for levels of affective involvement (Lester, Hoffman, & Brazelton, 1985; Gottman & Levenson, 1985; Tronick, Als, & Brazelton, 1980) or mood (Larsen & Kasimatis, 1990). Physiological time-series measures such as blood pressure or heart rate can be made in either social or nonsocial situations. Psychophysical judgment tasks, such as critical flicker fusion threshold, can be assessed over long periods of time to see whether thresholds vary systematically (Hammond, Warner & Fuld, 1995).

If the number of observations in the time series is very small (e.g., N ranging from 2 to 4), then panel analysis of survey data is probably the method of choice. If the number of observations is a little larger (N up to about 10), then repeated-measures analyses of variance (ANOVAs) or multivariate treatments of repeated measures may be appropriate. This book focuses on techniques that are useful for identification of patterns in much longer time series (N greater than 50 observations), and particularly methods for assessment of cycles that sometimes arise in long time series. Most time-series experts suggest that the use of time-series analysis requires at least 50 observations in the time series. Depending upon the length of cycles that are being examined, the number of observations in the time series may need to be even larger, as explained later.

The discussion in this book assumes some specific characteristics for time-series data. These characteristics are described in this section. For different types of time-series data (such as categorical time series) references are provided to sources that describe more appropriate statistical analyses.

Throughout this book, unless otherwise noted, the time-series data used in all examples are assumed to have these characteristics:

1. *The time-series variable is assumed to be a continuous variable that is (at least approximately) an interval ratio level of measurement.* Good examples of this are measures of systolic blood pressure or frequency counts of aggressive acts. These are examples of a true interval/ratio level of measurement, and so the statistical tools of time-series analysis, which use correlation and regression analyses, are appropriate.

Many researchers in psychology treat data such as Likert-scale ratings, which are probably closer to an ordinal than to an interval level of measurement, as if they were an interval/ratio level of measurement. That is, familiar parametric techniques such as correlation and regression are often applied to Likert-type rating data as a matter of convenience even though such data do not meet the formal measurement scale requirements for these analyses. This liberal approach to data analysis is used in this book. Categorical time-series data arise in many kinds of research, particularly studies that involve assessment of the occurrence of different behavioral states; but categorical data require different kinds of analysis than the ones covered in this book, such as log-linear models of serial dependence (see Gottman & Roy, 1990, Chaps. 10 and 11).

2. *Scores on the time-series variable should be approximately normally distributed.* Just as in correlation or regression analysis, non-normal distribution shapes or extreme outliers may require special handling. Outliers can be disproportionately influential on the conclusions of periodogram or spectral analysis, just as in regression or correlation analysis.

3. *Observations are obtained at equally spaced time intervals.* For instance, mood ratings could be obtained once per day, always at the same time of day. Unequally spaced observations or event-based rather than time-based sampling are other possibilities. However, for event-based time series, cycle length would need to be expressed in terms of number of events per cycle instead of number of time units per cycle (as in Van-Lear, 1991).

4. *Although there is no absolute rule about the minimum number of observations, the N of observations should be reasonably large. An N of at least*

50 is suggested here as a minimum. A few of the worked examples here use smaller *N*s to make it easier to enter the data, follow the analysis step by step, and perhaps even do some of the calculations by hand. However, this is done only for the sake of keeping examples relatively simple; longer time series should be used in actual research applications of time-series analysis.

Types of Research Questions about Time Series

Whether the observations in a time series represent homicides, mood ratings, physiological measures, or some other group or individual response, there are several basic kinds of questions about pattern in time series that can be asked about virtually any time series. These questions are described qualitatively in this chapter; later chapters show how quantitative indexes can be obtained (from trend analysis, harmonic analysis, periodogram analysis, spectral analysis, and related methods) to describe how well each of these types of pattern fits the observed time-series data.

It is helpful to begin this qualitative questioning about time series by visually examining graphs of some typical time-series data. Later chapters will show how we can obtain more quantitative and precise descriptions of the qualitative features that we see in visual examination. To begin learning what to look for, consider the following classic example, shown in Figure 1.1: the Wolfer sunspot data. This time series consists of the number of sunspots in each year from 1770 to 1869.

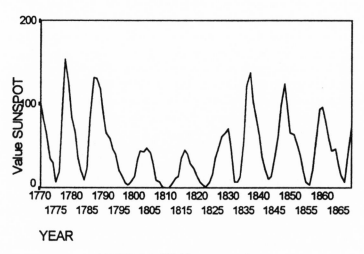

FIGURE 1.1. Wolfer sunspot data.

What can we learn from a visual examination of these data? What questions are suggested by this graph? First, it is obvious that the number of sunspots varies substantially over time: from a minimum of 0 to a maximum of 154. Second, it does not appear that there is a general increasing or decreasing trend over time. Third, peaks in number of sunspots tend to occur at about 11- or 12-year intervals. The spacing of these peaks and the height of these peaks is not perfectly regular, but it does appear that there are fairly regular cycles in sunspot activity. Finally, the amplitude or height of the peaks varies over time. The peaks are much lower during the middle years (from about 1795 to 1835) than in the earlier and later years. Possibly this is due to some other variable that influences sunspot activity over longer periods of time.

Spectral analysis and related analytic techniques make it possible for us to quantify these impressions about the time series. The analytic strategy outlined in this book involves the following steps:

First, it is a good idea to begin with data screening by employing the usual methods to assess distribution shape, outliers, and so on for the variable. We look at the number of sunspots as a continuous variable (ignoring the time factor), set up a histogram, assess whether the distribution of this variable is reasonably normal in shape, and decide whether any data transformation or special handling of outliers is required. In addition, at this point we might evaluate whether the amount of variation in sunspots is large enough to be of real interest. (If sunspots only varied in number from 4 to 5 from year to year, this might be too small a range to be of any practical importance.)

Second, we do linear and curvilinear trend analysis of the sunspot time series to see what percentage of variance in the time series is accounted for by trends, to test statistical significance of trend components, and to remove these trends before looking at possible cyclic patterns.

Next, we would apply periodogram analysis or spectral analysis (described in Chapters 5 and 6 of this book) to the *residuals* from this linear trend to see what percentage of the remaining variance in the time series (after trend removal) is due to cycles about 11 or 12 years long, as well as to determine whether there is evidence of any other cyclic pattern in the data. We can report the percentage of variance accounted for by major periodic components, and we can test for statistical significance to see whether the amount of variance associated with 11- to 12-year cycles is unlikely to be due to chance.

Finally, if we see irregularities in the cycles over time (such as the changes in the height of the peaks), we can do follow-up analyses such as complex demodulation (as described briefly in Chapter 7) to describe these changes in the nature (amplitude or height) of the cycles over time.

Other examples of pattern in graphs in time series can be found in later chapters. Figure 3.1 shows the numbers of airline passengers per month (in thousands) graphed across the years 1949 to 1960; this graph shows a strong linear trend, with 12-month cycles superimposed on it. Figure 6.1 shows measures of systolic blood pressure for a speaker in a conversation, taken once every 10 seconds; rather irregular cycles in blood pressure are evident, with sudden drops in blood pressure occurring at intervals of about 200–300 seconds.

In this book, analysis of time series will be treated as a relatively simple variance partitioning problem. We start with the overall variance of the time-series variable and break this down into several components to see which components account for the largest shares of variance. Figure 1.2 illustrates the logic of this variance partitioning using the sunspot time series as an example.

The first step is trend analysis: we divide the total variance of the time series into two components—the part that is accounted for by a linear (or curvilinear) trend, and the residual part that is not accounted for by such a trend.

The second step involves taking the residual variance (the variance not accounted for by a trend) and dividing it up into variance that can be accounted for by a set of cyclic components. The analysis that accomplishes this variance partitioning (periodogram or spectral analysis) is essentially a form of ANOVA that divides the total Sum of Squares for the entire time series into $N/2$ Sum of Square components. Each Sum of Squares component represents the part of the variance of the time series that is due to a cycle of a different period or length. The Fourier theorem

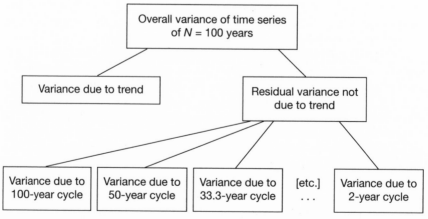

FIGURE 1.2. Logic of partitioning of variance for time series.

says that we can reconstruct any time series of length N by adding a set of cyclic components (sinusoids) with the following periods or cycle lengths: $N/1, N/2, N/3, \ldots, (N/2)/2$ or 2. This means, for instance, that a time series that is 100 years long can be reconstructed as a sum of cycles that are 100, 50, 33.3, 25, 20, ..., 2 years long. Each of these 50 ($N/2$, where N is 100) cyclic components has a Sum of Squares associated with it that has 2 degrees of freedom. Essentially, this is a form of ANOVA, in which the total Sum of Squares for the time series is divided into Sum of Squares components that are accounted for by each of the $N/2$ periodic components (see Box & Jenkins, 1970, pp. 36–39, for a more formal treatment and a fully worked empirical example).

If the time series is a set of random numbers, then the variance of the time series is approximately equally distributed across these 50 components, and the Sum of Squares that corresponds to each of the $N/2$ cyclic components is about the same. However, if the time series is clearly cyclical, then the fitted sinusoidal component that corresponds to the cycles that best fit the time series data accounts for a large share of the variance in the time series. So, for instance, when a periodogram or spectral analysis is performed on the sunspot data, the cyclic component that corresponds to cycles about 11 years long accounts for a much larger share of the variance than the shares of the other 49 cyclic components that are fitted to the time series.

The Sinusoid as a Model for Cycles

Although trend analysis can be handled by using methods and terminology that are already familiar to most behavioral or social scientists, analysis of cycles requires some new terminology and methods. More details (with worked examples) follow in later chapters; this section introduces some necessary terminology and ideas.

A set of closely related statistical methods have begun to be widely used to detect cycles in time-series data: harmonic analysis, periodogram analysis, and spectral analysis. These methods are described in more detail in Chapters 4, 5 and 6 of this book. In these analyses the model that is used to represent cycles is a sinusoid, that is, the waveform of the trigonometric sine or cosine function. This waveform is used because it is mathematically convenient and a reasonably good description of many naturally occurring cycles. If there are compelling reasons to model a waveform that has a different shape—such as a "boxcar" or rectangular waveform—then other methods not covered in this book, such as Walsh functions, may be more appropriate (see Broadbent & Maksik, 1992).

The family of all possible sinusoids can be generated by varying the four parameters of a sinusoid: the mean (μ), the period (τ), the phase (ϕ), and the amplitude (R). An example of a sinusoid with a μ of 10 and an R of 3 units on the Y axis, and a τ of 12 time units and a ϕ of 4 time units on the X axis, is shown in Figure 1.3 In harmonic analysis (see Chapter 4) the goal is to fit a sinusoidal waveform to an observed set of time-series data, estimating the period, mean, phase, and amplitude that make the waveform correspond most closely to the observed data. In periodogram analysis or spectral analysis, we will essentially be fitting a large set of different periodic components to see which one explains most of the variance in the time-series data. Sometimes several sinusoidal components need to be fitted to the time series in order to represent all the cyclic patterns in the time series, leaving residuals that are random.

Here is some basic terminology for the description of cycle parameters. The period of a sinusoid, τ, is the length of a cycle, that is, the distance from one peak to the next. In this book τ will generally be expressed either in number of observations or in number of units of time. When each observation corresponds to one time unit (mood measured

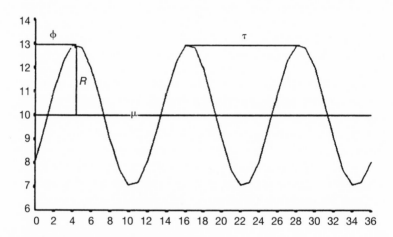

Equation that describes the sinusoid in the graph:

$$X_t = \mu + R \cos(2\pi t/\tau + \phi)$$
$$X_t = 10 + 3 \cos(2\pi t/12 + 4)$$

Parameters of this equation (example values):

μ = mean of sinusoid (10 units on Y axis)
R = amplitude of sinusoid (3 units on Y axis)
τ = period of sinusoid (12 units on X axis)
ϕ = phase of sinusoid (4 units on X axis)

FIGURE 1.3. Graph of a sinusoid function with a period of 12 observations, mean of 10 units, and amplitude of 3 units.

once per day; sunspots observed once per year), the number of observations and number of time units are equivalent and no conversion is needed. But if the sampling interval (Δt, the time between the measurements) is not one unit, it is necessary to convert period length estimate from number of observations to number of time units. For instance: if blood pressure is measured once every 10 seconds ($\Delta t = 10$) and the cycle that is detected has a period or cycle length of 20 observations, then this corresponds to a cycle that is $20 \cdot \Delta t = 20 \cdot 10 = 200$ seconds long. In most computer programs it is easiest to represent period in terms of number of observations per cycle and then convert this to an estimate of period in units of time by hand if necessary.

The frequency of a sinusoid, f, is the inverse of the period: $f = 1/\tau$. If a time series has a cycle that is 12 observations (12 months) long, then the corresponding frequency per time unit is 1/12 (i.e., 1/12th of a cycle occurs during one observation.) It is more conventional to talk about which "frequency" or "frequency bands" account for large shares of the variance in the data, rather than which "periods" account for large shares of the variance.

So far, frequency has been expressed in terms of number of cycles per observation. However, for most computations, frequencies are converted to radians. In practice, this simply means that the frequency in radians, often denoted ω, is obtained by multiplying f (or $1/\tau$) by the constant 2π.

The phase of a sinusoid, ϕ, is the location of the first peak (in a cosine function, for instance) relative to time zero. (If we are particularly interested in phase, then starting the observation count from 0 rather than 1 makes it a little easier to talk about phase; for this reason, in many of the examples that follow, the observations are numbered beginning with 0. However, if phase is not of great interest, or if the N in the time series is very large, it makes very little difference whether the observation count begins at time 0 or time 1.) Probably the easiest way to think about phase is as a time lag; for instance, in the sunspot data shown in Figure 1.1, the first peak occurs at time 0 (year 1770, the first year included in the time series-data). If we ignore this first peak, the next peak occurs at year 1778, a time lag of 9 years. However, in most computer printouts and in most computations, phase is expressed in other terms, such as: fractions of a cycle; degrees (where 360° is one full cycle); or radians (where 2π radians constitute one full cycle). Many times the investigator will find it useful to convert phase into a time lag, and this is easy as long as it is clear what units are involved. If the phase is π radians and the cycle length is 10 years, then notice that π is half of the 2π total cycle length. Therefore π radians would correspond to one-half a cycle: for a cycle of 10 years, this implies a phase of 5 years.

The amplitude, R, is the height of the peak relative to the mean of the waveform. This is generally expressed in the units that the dependent variable is measured in, for example, number of sunspots; millimeters of mercury (mmHg) in blood pressure. See Figure 1.3 for a graphical illustration of these parameters.

When we fit a sinusoid to time-series data, we are essentially asking whether peaks or troughs tend to occur at regular time intervals in the time series. Implicitly we are assuming that the shape of cycles in the data is at least approximately sinusoidal. This may not be an accurate assumption in many situations; for instance, consider the sunspot time-series data shown in Figure 1.1. The cycles that are apparent in these data have sharper peaks and flatter bottoms than an ideal sinusoid. However, sinusoids are a mathematically convenient approach to modeling cycles, and so they are widely used even in situations (such as the sunspot data) where the shape of the cycles really is not sinusoidal. In some cases, fairly complex waveform shapes can be modeled using a composite of several sinusoidal components.

If it is important to have more accurate information about the shapes of cycles in the data, or if the shapes of cycles deviate extremely from the ideal sinusoid, then a data analyst may need to turn to other less common and somewhat less mathematically tractable methods of analysis. (These are beyond the scope of this book.) I will assume that the time-series data in all subsequent examples are close enough to sinusoidal for this method of analysis to be a reasonable approximation.

Examples of Empirical Research Involving Cycles

The statistical methods covered in this book provide methods for quantitative assessment of these two kinds of predictable pattern in time-series data: trends (linear or curvilinear) and cycles. These statistics can be used to answer questions such as the following: Are there statistically significant trends or cycles in the data? How much of the variance of the time series is accounted for by trends and cycles? What are the parameters for these patterns (i.e., the intercept and slope of the trend line; the period, amplitude, and phase of the cycles)?

Why might such questions about patterning over time be of interest to social psychologists? There is a growing body of theoretical and empirical work suggesting that statistical patterning and contingency of behavior are related to people's perceptions and evaluations of responsiveness, predictability, and rapport (Warner, 1992c). In other words, *qualitative* evaluations of social interaction (such as how responsive I feel my conversation partner is to me) may be related to *quantitative* indexes of patterning in behaviors (such as whether my partner's vocal activity is closely coordinated with my vocal activity).

Questions about patterns in time series, such as trends and cycles, arise in many different research situations. For instance, in analysis of economic time-series data (such as records of sales), it is often of interest to assess whether there are overall increasing or decreasing trends and also whether there are weekly, monthly, seasonal, or even longer cycles. Often in econometric analyses trend and cycles are viewed as sources of artifact that must be removed from the time-series data prior to doing other analyses. For instance, before assessing the relation between cotton prices and lynchings (Hovland & Sears, 1940; Hepworth & West, 1988), the analysts removed trends from each time-series variable. In econometrics, most of the literature on trends and cycles focuses on the issue of how to remove these from time-series data so that other analyses (such as examination of relationships between the trend residuals for the two time series) can be performed.

In other research domains, such as research on biological rhythms, researchers are interested in assessing cycles in physiological functioning that are various different lengths: circadian (about 24 hours), ultradian (less than 24 hours), monthly, or longer cycles. For these researchers, cycles are a subject of interest rather than a potential source of artifact. Biological rhythms researchers tend to be interested in *describing* cycles rather than *removing* them from the data. In this book I have adopted a perspective that is closer to that of the biologists: I am more interested in *describing* cycles as an interesting feature of the data than in *removing* them from the data. The difference between the econometric and the biological rhythms approaches is not a difference in the analysis: the same statistical procedures are sometimes used in both research domains. The difference is mainly in the way results are reported and what information is included. The components of the time-series data that econometricians tend to remove, and sometimes even dismiss as artifactual (trends and cycles), are the features of the times-series data that biological rhythms researchers focus on as interesting and important.

Chapter Summary and Plan of the Book

This chapter was a relatively nontechnical introduction to some basic issues and vocabulary that will be used in later chapters. First, the types of data most often used in time-series research were described. Second, the sinusoid was introduced as a model for "cycles" in time series. Third, an overview of a suggested strategy for data analysis involving a description of trends and cycles was provided.

Although the mechanics of data recording and the theories that are being tested differ greatly across these research domains, many researchers share a common interest: describing patterns over time (trends and cycles) in time-series data. The first half of this book outlines a process for examining *univariate* time-series data with a systematic set of questions in mind (listed below). Basically, no matter what the research domain (psychology, physiology, or social indicator data), these questions may be a useful way to structure the inquiry. The second half of the book outlines suggested procedures to handle *bivariate* time-series data, for instance, data on blood pressure and amount of talk collected concurrently. Finally, other issues will be discussed, including ways of summarizing results across a large number of time series, and ways of describing changes in the behavior of a time series over time.

The following questions about univariate time series are addressed in upcoming chapters (Chapters 3 through 7):

1. Does the time series show a clear linear or curvilinear trend? What percentage of the variance in the original time series is accounted for by the trend component? (If there are trends, then subsequent analyses are usually performed on residuals from this fitted trend).

2. Do the trend residuals show one (or more) major periodic components or cycles? If yes: What periods/frequencies account for most of the variance? What percentage of the variance in the trend residuals (and in the original time series) is accounted for by these cyclic components? These questions can be answered using the related statistical methods that are described in subsequent chapters of this book: harmonic analysis, periodogram analysis, and spectral analysis.

3. Going back to the original raw time series: How well do the combined trend and cyclic components reconstruct the pattern in the time series data? Are the residuals "white noise" or is there evidence of further pattern?

Later in the book there will be comments on ways of describing changes in the behavior of the time series over time, for instance, changes in the amplitude of cycles. As a result of the univariate analysis, the original time-series data can be represented as a sum of several (mutually uncorrelated) component time series: the trend, one or more cyclic components, and residuals. The details of this representation will be described in Chapters 3 through 6.

CHAPTER 2

Issues in Time-Series Research Design, Data Collection, and Data Entry: Getting Started

Introduction

This chapter outlines basic design decisions that a researcher needs to make before undertaking time-series research. Some of these design issues are similar to those in other types of research (such as subject selection and control over extraneous environmental variables). Discussion of these factors highlights any special considerations in making these decisions in the context of time-series research. Other design decisions (such as determining the length of the time-series data record and the sampling frequency) are unique to time-series research, and it is useful for the researcher to try to anticipate the types of pattern that he or she hopes to find in the data before making these decisions.

All the research and data analyses described in this book involve nonexperimental or descriptive types of research, as there are no manipulated independent variables. (For an introduction to analysis of interrupted time-series experiments, see McCleary & Hay, 1980.) Even in descriptive or nonexperimental research, however, it is important to control for extraneous variables that might influence the behavior being observed, in order to limit the number of explanations that have to be taken into account when trying to explain the results of the study. Usually in nonexperimental time-series research, the investigator hopes to describe some naturally occurring process over time. This means that data should be collected in ways that make it unlikely that artifacts will produce spurious patterns in the data. A review of "threats to internal validity" in quasi-experimental research will be presented as a reminder of some of the general types of extraneous variables that need to be considered, and controlled if possible, when collecting time-series data.

The typical goal of a time-series study is to describe some kind of naturally occurring pattern in behavior over time. Fiske and Rice (1955) discussed several of the factors that can give rise to patterned intraindividual response variability over time, for instance, adaptation, practice effects, or fatigue. Campbell and Stanley (1966, pp. 5–6) used the term "maturation" to include many kinds of naturally occurring variability in time-series data. The term "maturation" refers to "processes within the respondents operating as a function of the passage of time per se (not specific to the particular events) including growing older, growing hungrier, growing more tired, adaptation to the situation, practice effects, and the like" (Campbell & Stanley, 1966, p. 5).

In the more familiar context of an experiment intended to assess the impact of an intervention such as a pretest–posttest design or an interrupted time-series experiment, "maturation" is viewed as a threat to internal validity. That is, maturation competes with the manipulated independent variable as a possible explanation for any changes that are observed over time. In this book, however, the kinds of intraindividual response variability that Campbell and Stanley included in their broad definition of maturation are the very factors that we want to examine when we look at naturally occurring response variability over time in nonexperimental time-series data. Statistical methods such as spectral analysis can be used to assess patterning in intraindividual responses over time. Naturally occurring response variations over time, such as daily or monthly cycles in mood or in internal physiological states, are the focus of interest in spectral analysis studies. (Of course, changes in the amplitude or period of cycles could also be examined within the context of an experiment; a manipulation might change the amplitude of cycles, for example.)

Design of a time-series study involves many decisions that are familiar to researchers, such as selection of subjects, setting, and variables to observe. It also involves decisions about issues that may be less familiar (such as sampling frequency). The researcher's fundamental design decisions include those discussed in the following sections.

Selection of Subjects

What types and numbers of subjects should the researcher include in the study? As in other kinds of research, time-series researchers typically hope that their sample of subjects is representative of some larger population of interest. However, the logistics of collecting time-series data are often restrictive. For example, when behavioral or physiological time-series data are collected, unusual amounts of time or cooperation from research subjects may be needed and it may only be possible to collect extensive time-series data for a small number of subjects.

Dukes (1965) pointed out, in his classic paper on $N = 1$ research, that psychological research has tended to include a large N (of subjects) but often has included a small N, or an N of 1, with respect to other factors in the study (such as setting, time period, on experimenter). It is worth noting that including only $N = 1$ time points in a study limits the generalizability of results to other time periods and makes it impossible to assess stability or change in behavior over time. Some time-series studies have used $N = 1$ or a small number of subjects in order to look at a large number of time periods. When only one time series is available, researchers are limited to using the data in hand. Of course it is desirable to have a large number of time series, so that any pattern that is detected in one time series can be evaluated to see if it occurs in other time series as well. Chapter 7 discusses ways of summarizing results when many time series have been analyzed.

Note also that the "subject" for which time-series data are available can be an individual organism (a person or an animal) or it can be a social group, small or large (as in social indicator data that consist of rates of suicide, homicide, or other behaviors for cities, states, or countries).

Setting and Environmental Conditions

Where and under what conditions shall the time-series data be obtained? How can potential contamination of time-series data by environmental events or experimenter expectancy be most effectively minimized? Biologists have often been interested in "endogenous" factors or internal regulatory processes in organisms that may give rise to cycles in behavior and physiology. That is, they are interested in the possibility that the cycles primarily arise from regulatory processes internal to the organism (although these cycles are also modulated by environmental events). In order to assess what types of pattern arise from endogenous regulatory processes, biologists make observations of behavior or physiology under "free-running" conditions. The environment is kept as constant as possible, so that environmental variations such as the 24-hour alternation between light and dark periods are eliminated. This makes it possible to assess the extent to which observed rhythms are due to endogenous regulatory processes. Once this "baseline" for the naturally occurring processes has been established, later research can assess how these naturally occurring cycles are modulated by environmental events such as the light–dark cycle, variations in temperature, availability of food, and even the presence of other organisms (Aschoff, 1981). Trying to hold environmental conditions as constant as is practicable is an attempt to control for the class of threats to internal validity that Campbell and Stanley (1966) called "history": events that occur between measurements that might have affected the behavior.

Choice of Measurement Method and Problems with Measurement Artifact

Will the measurements be based on self-report or behavioral observation? What type of mood-rating items or behavior coding system will be used? Time-series analysis can be performed on almost any type of data where repeated and regular measurements can be obtained at reasonably low cost, such as physiological monitoring, and rating or coding of observed behaviors. However, even when automated systems are used to make time-series measurements, researchers need to be aware of common problems that can compromise the quality of the data.

Campbell and Stanley (1966) noted two types of measurement artifacts that can be threats to internal validity in pretest–posttest or any other time-series designs: "testing" and "instrumentation."

"Testing" refers to the problem that the very process of measurement sometimes changes the behavior or mental state that is being measured, and measuring repeatedly is a requirement of time-series designs. For instance, when people fill out daily symptom checklists, over time they become sensitized to physical symptoms and tend to report more and more symptoms. Attaching electrodes or sensors to subjects may increase their physiological arousal. It is often necessary to allow adaptation periods at the beginning of time-series studies to allow the subjects responses to return to baseline and to allow any initial reactivity to measurement to subside.

"Instrumentation" is also a potential problem, whether the measurement is done by a human observer or by an automated system. Suppose that the time-series data consist of ratings of emotional involvement made by a human observer who is watching a videotape of a social interaction. As the rater changes over time he or she may become more expert, more bored, more fatigued, and the like, and any of these changes in the rater could produce artifactual changes over time in the time-series record. Even an automated system can show drifts or changes in calibration over time. The resistance of an electrode may change over time, or changes in a subject's posture can affect the readings that are obtained. Whether the measuring "instrument" is a human observer or some automated system, it is necessary to collect calibration data often enough to make sure that the measurements are consistent over time.

A trend in a time series might represent a "testing" problem (as the subject's responses change as a function of being repeatedly observed or tested) or an "instrumentation" problem (as the calibration of the measuring device drifts over time). A researcher cannot confidently attribute patterns in time-series data to naturally occurring endogenous cycles unless these potential sources of artifact can be ruled out. Baseline periods for adaptation to the observation situation, or practice sessions, may be needed to get rid of artifactual start-up patterns. Frequent calibration checks are needed (on either human observers or automated systems) to make certain that the measurement procedure is not systematically changing over time.

Other Threats to Internal Validity

"Statistical regression" can be seen in time-series data; any extreme outlier observation will tend to be followed by observations closer to the mean, as an artifact of regression toward the mean. Statistical regression is most likely to be a source of artifact when the monitoring of subjects is initiated after the occurrence of unusually high or low levels on the variable of interest. This sometimes happens in studies of archival data on social indicator variables, as policy makers tend to become most interested in the crime rate and other indexes of social pathology when these are unusually high.

"Experimental mortality" can be a major problem in time-series research. Some subjects may drop out of the observation earlier than others for a variety of reasons. For instance, subjects who do not complete the data collection process may well be systematically different (in cooperativeness and perhaps in other ways as well) from subjects who do.

Thus, Campbell and Stanley's (1966) list of threats to internal validity is a useful guide to other variables (besides naturally occurring regulatory processes within individuals) that could create patterns in time-series data. If we are interested in describing naturally occurring patterns, then these other kinds of variables ("history," "testing," "instrumentation," and so forth) are potential sources of artifacts that we need to take into account before we attribute any patterns in our data to natural developmental processes. Controlling for these and other threats to validity is as important in nonexperimental time-series studies as in more traditional types of experiments.

Basic decisions about the type of subjects, the setting, and the methods of measurement can be made more intelligently if Campbell and Stanley's excellent advice on factors that can affect internal and external validity is thoughtfully applied. However, there are some additional design issues that are specific to time-series studies. These are addressed in the remainder of this chapter.

Determination of Time-Series Length

The most important issue to consider when deciding on an appropriate length (number of observations) for the time-series study is the length of the longest cycle that the researcher is interested in detecting and describing. For example, if a researcher measures mood once per day and is interested in assessing 7-day cycles in mood, then the absolute minimum time-series length would be $N = 7$ days. It would be much better to obtain a time-series record long enough to include several repetitions of the cycle. For reasons that will become clear later (in Chapter 5, when "leakage" is discussed), it is most convenient if the length of the time series is an integer multiple of the cycle length that is of primary interest. It is also a good idea to have at least 5 or preferably 10 repetitions of the cycle included in the data record. Thus, if the researcher is interested in 7-day cycles in mood, it would be a good idea to obtain a data record at least 35 or preferably 70 days in length ($5 \cdot 7$ or $10 \cdot 7$).

A second consideration is a purely pragmatic one: how long a time series is it reasonable to obtain? Some kinds of time-series data collection require subjects to sit quietly in controlled laboratory conditions, and there may be a limit to the length of time that people can be persuaded to do this. When archival time-series data are used, researchers often must use the data that are available even if the time-series record is briefer than might be desirable.

A third (usually less important) issue to consider when deciding the number of observations in the time series is whether the computer program being used has any special requirements or limitations about the number of observations. The SPSS TRENDS SPECTRA program requires that N (number of observations in the time series) be an even number; if the N for the input data is an odd number, the last observation is dropped. A few older computer programs use algorithms that require N to be a power of 2, because they use special computational shortcuts that work only in this case. Most programs now provide more flexibility in choosing N.

When periodogram analysis or spectral analysis are performed on time-series data, the longest cycle that is fitted to the time series is one that is the same length as the time series. The shortest cycle that is fitted to the data is determined by the sampling frequency, which is discussed in the next section.

Selection of Sampling Frequency

Together, the selection of sampling frequency and time-series length determine the set of frequencies (or cycle lengths) that can be detected in the time-series data. Sampling frequency is the time interval that corresponds to one observation (Δt). For instance, if mood is assessed once per day, then Δt = 1 day; if blood pressure is measured every 10 seconds, then Δt = 10 seconds. Observations can be made once per second, once every 10 seconds, once a minute, once a day, and so forth. The shortest cycle that can be detected corresponds to a cycle two observations long. The longest possible cycle that can be detected is N observations long, where N is the length of the time series. If mood is measured once per day for 28 days, then the shortest possible cycle in mood that can be detected is 2 days and the longest cycle that can be detected is 28 days.

Another issue involved in deciding on a sampling frequency is the lead–lag interval that might be important in bivariate time series. For instance, if the researcher thinks that a mother might respond to an infant's change in behavior within 1 second, then he or she should sample at least once per second (or even more often) in order to detect this lead–lag relationship.

The choice of an inappropriate sampling frequency can lead to a type of artifact known as "aliasing," which can be explained in a nontechnical way through an example, shown in Figure 2.1. (A technical explanation is provided by Bloomfield, 1976, pp. 26–27.) Suppose that the variable that is being studied shows fairly rapid cycles (shown by the

FIGURE 2.1. Illustration of "aliasing."

dark line in Figure 2.1) but it is sampled at a frequency much slower than two times per cycle. The obtained time-series data, illustrated by the light line in Figure 2.1, suggest a much lower frequency or longer cycle in the actual underlying continuous process. The spectrum or periodogram computed for this obtained data will have a peak corresponding to this "long" cycle. This phenomenon is called "aliasing" because the higher frequency "masquerades" as a lower frequency in the observed data. The remedy for this problem is to make sure that samples are taken frequently enough to capture the highest frequency present in the data: at least two times per cycle—or, ideally, even more.

It is not possible to judge whether aliasing is a problem simply by looking at the observed data. To determine whether aliasing is occurring, it is necessary to know what happens if samples were obtained at much higher frequencies; this would reveal the shorter cycles if present. In practice this means that the researcher needs to know something about the frequency composition of the signal that is being sampled.

If in doubt, initially the researcher should sample at the highest frequency that is feasible, in order to assess whether higher frequencies are present. If high frequencies are present, these can be filtered out or removed from the signal either by electronic filtering devices, such as polygraph filtering modules, or by sampling at a very high frequency and then averaging together or smoothing the observations to get a time series with fewer observations.

To summarize: Here are some suggested guidelines for determination of sampling frequency (Δt) and time-series length (N). The sampling frequency or interval between observations should be brief enough so that the shortest cycle you might be interested in detecting would correspond to an absolute minimum of two observations; ideally, the researcher would want to have a larger number than two observations for each cycle in order to get a better idea of what the shape of the cycle is. For instance, if the researcher is interested in the possible existence of cycles that are 10 seconds long, he or she could take one observation every 5 seconds (however, this would be the absolute minimum; it would be preferable to take one observation per second, so that there would be 10 observations per cycle). If the researcher is in doubt about whether aliasing might be a problem, or if he or she is not certain what the shortest cycle length might be, it would be better to sample more often (at smaller time intervals). Data can always be aggregated into larger time blocks later, if that proves convenient. The length of the time series (number of observations) should at an absolute minimum be the same as the cycle length of the largest cycle that the researcher is interested in detecting; but it is preferable to observe several repetitions of the cycle that is of primary interest, not just one repetition. A longer time series makes it possible to see whether the cycles actually repeat regularly.

Time-Based versus Event-Based Records

Most of the examples described in this book assume that the time series consists of observations made at equally spaced time intervals, such as 1 second apart. However, some time-series records are event based rather than time based. If a researcher records the amount of self-disclosure that occurs during each of 30 conversations as a time series, then there will be 30 data points that correspond to events rather than equally spaced time intervals. In principle, there is no reason why one cannot apply the methods described here (harmonic analysis, Chapter 4; periodogram analysis, Chapter 5; spectral analysis, Chapter 6) to an event-based time series. However, the statements about the length of cycles would then be made in terms of number of events instead of units of time. In practice, it may be possible to estimate how much time corresponds on average to each event and to estimate the length of each cycle in time units.

Human versus Automated Systems Coding

The examples mentioned here mainly include situations where the time-series variable is recorded by an automated system (e.g., a measure of systolic blood pressure made every 2 seconds by a noninvasive blood pressure monitor). However, the same statistical methods can be used in research where the time-series variable consists of ratings or judgments made by trained observers (such as Tronick and coworkers' [1980] monadic phases). Automated systems have some potential advantages: They may be more reliable or consistent, and less labor intensive and expensive, than many human observer coding methods. However, there are qualitative characteristics of social interaction that it may be impossible to assess using any existing automated system (such as intimacy, or positive/negative affect), and so there are many cases where human observers will be the source of the time-series data. Whether the data come from a human coder or an automated system, issues such as reliability, validity, and reactivity have to be taken into account in evaluating the quality of the data. It is not safe to assume, simply because a data collection system is automated and employs sophisticated technology, that it necessarily generates reliable or valid data; assessment of reliability and validity remain important even when automated systems are used for collection of time-series data.

Categorical versus Continuous Measures: Levels of Measurement

Most time-series data can be classified into one of the following four types:

Type a. *Categorical data* (with multiple categories). Each observation consists of a code that labels one of several possible behavioral states. For instance, in peer play, 1 = plays alone, 2 = plays with another child, 3 = interacts with teacher, 4 = sits and does nothing, etc.

Type b. *Dichotomous data* (categorical data with only two codes). For instance, a string of 1's and 0's can be used to represent the presence and absence of some action. Jaffe and Feldstein (1970) used "1" to stand for "talk" and "0" for "silence"; the on–off pattern of vocalization and pauses can then be represented by a string of 1's and 0's. Similarly "1" could indicate "gazes toward partner" and "0" could indicate "gazes away from partner."

Type c. *Data that are continuous but that are not a true interval/ratio level of measurement.* For example, variables such as mood or attitude are often assessed by means of ratings or assessments (such as Likert-type scales, or dial settings). These measurements probably are not true interval/ratio data, but it is a fairly widespread practice in data analysis to use ratings and similar data as if they were a good enough approximation to interval/ratio data to justify the use of parametric statistics such as Pearson's r.

Type d. *Continuous data that are a true interval/ratio level of measurement.* In this case, each time-series observation represents an interval/ratio type of measurement, for example, the blood pressure reading given in millimeters of mercury (mmHg), the percent of time spent talking, or the interpersonal distance measured in centimeters.

According to the strictest standards, only data type "d"—true interval/ratio continuous data—would be considered appropriate for periodogram analysis and spectral analysis. Strictly speaking, spectral analysis (like other parametric statistical methods) should only be applied when the data are a true interval/ratio level of measurement. However, I will use parametric statistics with data type "c," such as Likert-type mood ratings, even though this type of data does not satisfy the strictest measurement requirements for the use of parametric procedures. This liberal practice (using parametric statistics with rating data that are not a true interval/ratio level of measurement) is common. A spectral analysis of, for instance, mood-rating data seems reasonable, even though mood ratings are not true interval/ratio data.

In addition, some of the methods for cycle detection that are described here can be applied to dichotomous time-series data (data type "b") with only minor modifications (see Gottman, 1981a, for details).

Type "a" time-series data, categorical time-series data with more than two possible categories in the coding system, are *not* suitable for harmonic or spectral analysis, however. Other methods of analysis should be used with such data; these are beyond the scope of this book. Appropriate methods for the analysis of serial dependence in categorical data, such as log-linear analysis, are presented by Gottman and Roy (1990).

Other Approaches to Describing Pattern in Time-Series Data

It is possible to consider other kinds of patterns in time-series data (apart from trends and cycles). There are many ways to describe systematic or predictable patterning in a single time series over time. From the 1940s until about 1980, a popular approach was the application of Markov chain analysis to categorical time-series data on behavior states. Outstanding examples include work by Jaffe and Feldstein (1970) and Bakeman and Brown (1977). Lagged conditional probabilities were computed to assess whether current behavior state could be predicted from one or more past behavioral states at levels significantly higher than chance. Analysts who have used lagged conditional probabilities and Markov models to look at patterns in on–off vocal activity, in gaze, or in transitions among behavior states have tended to find that behaviors were sequenced nonrandomly; many found that low-order Markov process models (i.e., models that predict the current behavior state from one or two previous behavioral states) were adequate to model the serial dependence in their data. An excellent tutorial introduction to this type of modeling was given by Jaffe and Feldstein (1970), and this approach is still extremely useful when the time-series data are categorical, that is, a sequence of different behavior state codes. Currently, many researchers use log-linear analysis to examine lagged dependence in categorical time-series data (Gottman & Roy, 1990).

Chapter Summary

This chapter outlined design decisions that need to be made prior to collection of time-series data. Some of these decisions (such as choice of subjects, setting, control for extraneous environmental variables, selection of types of measures, or controlling for measurement artifacts) are familiar because the same issues arise in many other types of research. A few of the design issues (length of the time series, sampling frequency, or event versus time based sampling) are unique to time-series research.

Basically, the researcher needs to consider two types of questions before designing the study. First, what types of pattern in the time series is the researcher interested in detecting? If the researcher wants to be able to see 7-day cycles in mood, then this will influence decisions about the sampling frequency and the length of the time series. If the researcher wants to detect lagged responses of a mother to her infant's behavior and the time lag is believed to be on the order of 1 second, then this will influence the choice of sampling frequency. Second, beyond description of these cycles, what kinds of attributions does the researcher want to be able to make about the factors that cause or influence these cycles? The analyses described in this book assume nonexperimental designs, and therefore causal inferences are never warranted. However, at a minimum, the researcher may want to design the study in ways that will make it possible to rule out at least some potential artifactual explanations for any cyclic patterns that are detected—namely, "history," "instrumentation," "testing," and other sources of problems.

Appendix 2.1. Entering and Graphing Time-Series Data Using Three Statistical Packages

SPSS Running under Windows

Readers who are unfamiliar with SPSS for Windows and wish to use this program for spectral analysis may find it helpful to duplicate this example by entering the data and going through the operations, as a way of getting acquainted with SPSS for Windows. This small data set (hypothetical mood data) provides an uncomplicated and brief example. Real datasets will typically have many more observations and will require more complex analysis than this first simple example.

The mood data introduced here will be used again in Chapter 4 on harmonic analysis and Chapter 5 on periodogram analysis. This example presents simulated data, similar to data one might obtain from a study of mood over time. Suppose that one subject rates her mood on an analogue scale once each day by putting a mark on a line ranging from 1 (very negative mood) to 6 (very positive mood). The researcher measures the position of this mark and reports it to the hundredths decimal place. Data are collected on a daily basis for 14 days. Because it is sometimes more convenient to use a time counter that begins with 0 (particularly when one is talking about phase), the day number will be given starting at 0 rather than at 1.

The data entry and graphics shown in this example make use of the SPSS Base System program documented in Norusis (1993). Later chapters also use several programs in the SPSS TRENDS program, which is sold as an add-on to the base package. To create an SPSS for Windows worksheet that contains these data (as shown in Table 2.1), the user would need to do the following. Note that any command shown in parentheses (   ) indicates a menu option that should be selected by clicking the mouse on it. Any word shown in quotation marks represents a label or name that the user types in (do not type the quotation marks).

Step 1. Start the SPSS for Windows program by double clicking on the SPSS icon in the Windows environment. An empty data worksheet will appear on the screen. The raw data will be entered in columns 1 and 2 of this worksheet. Select the first column of this data worksheet by clicking the mouse when you are pointing at any cell in column 1.

Step 2. Select the (Data) (Define Variables) commands from the menu at the top of the screen. Enter the variable name for the first column. In this example the column 1 variable will be called "day." If desired, click on the Change Settings: (Type) box to specify a different number of decimal places than those displayed by the default. The default is to display 2 decimal places, but in this example 0 decimal places were chosen for the first variable because the variable

TABLE 2.1. SPSS Worksheet "mood.sav" Containing Mood Time-Series Data

	Day	Mood
1	0	4.99
2	1	3.37
3	2	3.06
4	3	2.41
5	4	2.91
6	5	3.69
7	6	4.88
8	7	5.35
9	8	2.37
10	9	2.59
11	10	2.75
12	11	2.42
13	12	3.51
14	13	4.70

"day" is given in whole numbers. Click on (OK) when the variable definition is the way you want it.

Step 3. Click on column 2 of the data worksheet to select it, and go through the same process to assign the variable name of "mood" to column 2.

Step 4. Now you are ready to type in the data values. On each line, you type the day and the mood rating, copying the values shown in the sample SPSS worksheet in Table 2.1. When you have finished this step your SPSS worksheet should look like Table 2.1.

Step 5. When the values are all entered, select (File) and (Save As) to save your SPSS worksheet as a disk file. Type in a file name, including a subdirectory name if applicable. In general, SPSS data sets have the extension ".sav." In this example, the name of the file is "mood.sav." If the file is located in the subdirectory "\spsswin" on disk C:, its full name will be "C:\spsswin\mood.sav."

Step 6. To generate a graph of this time series with "day" on the X axis and "mood" on the Y axis, select (Graphs) (Line) from the menu at the top of the screen. Three options for type of graph are given: the default type, (Simple) is used here. Under the heading *(Data in chart are)*: change the selection from the default (summaries for groups of cases) to (values of individual cases).

Step 6. Select (Define). The screen now contains a pull down list of the variables in your file ("day" and "mood"). To specify your dependent variable (plotted on the Y axis) as the mood ratings, click on (mood) in the variable list. To move this variable into the box that is labeled (Line represents), you click on the arrow that appears on the screen and points to the dependent variable box. Mood will now be moved into the box for (Line represents). By default, the X axis will be labeled using the observation number. If you want the X axis to be labeled by day, then you click on (day) in the list of variables, then click on the arrow on the screen to move the variable (day) into the box labeled (Category Labels).

Step 7. When you have the graph specified the way you want it, select (OK). After a brief delay, the graph will appear in a new window, superimposed on the data worksheet window. It should look like the graph in Figure 2.2.

Step 8. To print this graph, select (File) (Print) (OK) from the menu at the top of the screen.

Step 9. If you want to modify this graph, select the (Edit) option from the pair of options—(Edit) (Discard)—appearing at the top of the graphics display window. If you are finished looking at this graph, you may save it in a file by selecting (File) (Save). Graphics files are usually saved with a file name that includes the extension "cht." If you want to discard this graph and go back to the data worksheet window, select (Discard).

Step 10. To finish the SPSS session, select (File) (Exit).

In future sessions, to retrieve the worksheet containing the simulated mood data, select (File) (Open) (Data). Then either type in the file name ("mood.sav") or select it from the pull-down list of data file names. The simulated mood data will be used as an example when harmonic analysis and periodogram analysis are introduced in Chapters 4 and 5.

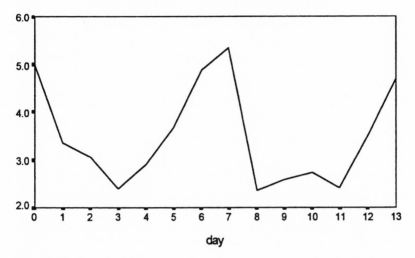

FIGURE 2.2. Graph of mood time-series data generated by SPSS for Windows.

BMDP 1T (BMDP/Dynamic Version 7.0, Running under DOS)

Step 1. Using a text editor, create a data file in which each line has the value of day and mood entered, as in the sample file MOOD.DAT shown below.

0	4.99
1	3.37
2	3.06
3	2.41
4	2.91
5	3.69
6	4.88
7	5.35
8	2.37
9	2.59
10	2.75
11	2.42
12	3.51
13	4.70

Step 2. Start up BMDP/Dynamic. If you are running it from DOS, and the programs are located in a directory called /dynamic, these commands will start up the program:

```
cd   /dynamic
bmdpdyn
```

Step 3. From the menu, choose EDIT. In the blank screen that appears, type in the commands as shown in the sample file MOOD.INP. Note that each new command begins with a "/" (solidus) and that many commands end with a "." (period).

/ INPUT VARIABLES ARE 2.
FORMAT IS FREE.
FILE IS 'MOOD.DAT'.
/ VARIABLE NAMES ARE DAY, MOOD.
/ END
SNAPSHOT VARIABLE = MOOD./

Step 4. Select "Run" by hitting the [f6] key.

Step 5. From the menu of BMDP programs that appears in the pull-down menu, select the program that you want to run, move the cursor to highlight it, and press [enter]. The program used in this example is 1T, "Univariate and Bivariate Spectral Analysis."

Step 6. The output file now appears on the screen. To step through it page by page, keep hitting [enter]. The output file can be printed and/or saved by selecting these items from the menu at the bottom of the screen.

Note that the graph of the mood time-series data appears as points, rather than as a connected line (Figure 2.3). The SNAPSHOT command tends to pro-

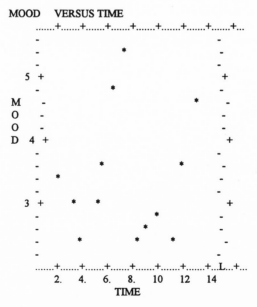

FIGURE 2.3. Output of mood time-series data from BMDP/Dynamic.

TABLE 2.2. SAS Command File, "mood.sas"

```
                goptions device = tek4015
                cback = stb
                gunit = pct
                htitle = 6
                htext = 4
                ftext = complex
                ctext = yellow;
data mood;
        input day mood;
        cards;
    0   4.99
    1   3.37
    2   3.06
    3   2.41
    4   2.91
    5   3.69
    6   4.88
    7   5.35
    8   2.37
    9   2.59
   10   2.75
   11   2.42
   12   3.51
   13   4.70
;
title1 'Figure 2.4';
title2 'Simulated Mood Data';
proc gplot    data = mood;
        axis1   value = (f = simplex)
                color = yellow
                label = (f = simplex a = 90 'mood rating')
                order = 1 to 6 by 1;
        axis2   value = (f = simplex)
                color = yellow
                label = (f = simplex 'day')
                order = 0 to 16 by 2 ;
        plot mood * day = 1 /
            vaxis = axis1
            haxis = axis2;
        symbol1 C = red I = splines;
        run;
```

duce a more intelligible graph if the time series is relatively brief. The SNAP-SHOT plot of the time series is displayed with time on the horizontal axis on the computer screen. If the time series is long, the TPLOT command (output for this is not included here) may be a better choice. When using TPLOT, time is plotted as the vertical axis on the graph.

SAS Running under UNIX

Using a text editor, create a SAS command file. The file "mood.sas," shown in Table 2.2, illustrates a relatively minimal set of commands to generate a graph in SAS (Figure 2.4). In this example, the data are typed into the same file as the commands, but data can be placed in a separate file.

Many of these SAS graphics commands are hardware specific. For this reason, the SAS commands shown here will probably require modification to work on your system; this example is merely suggestive, and the user will need to refer to SAS/GRAPH and other SAS manuals for full documentation in order to determine exactly what commands are needed on the user's own system. For instance, in the "goptions" command, the sample command file I am using includes the subcommand "device = tek4015." This means that the terminal on which graphics are being displayed is emulating a Tektronics 4015. You may need to replace this with a different type of device that corresponds to your own terminal's graphics mode.

In this example, other commands specify the colors of elements of the

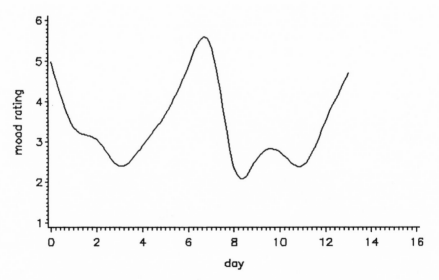

FIGURE 2.4. Graph of simulated mood data generated by SAS.

graph (background = stb, or standard blue; axes and labels are yellow; data points and interpolated line are red). The type of font used in the title text is specified as complex. The range of values on the X and Y axes are specified in the "order" subcommands: the Y axis values range from 1 to 6; the X axis values range from 0 to 16. The subcommand "*I* = splines" tells SAS to connect the discrete data points with interpolated splines, that is, relatively smooth curves.

CHAPTER 3

Preliminary Examination of Time-Series Data

Introduction

This chapter deals primarily with the kinds of information that a researcher can obtain about a time series from examination of graphs and from application of familiar simple statistical analyses. The procedures outlined in the following sections include useful exploratory and preliminary techniques. Based on the preliminary examination of the time-series data, the researcher can make several decisions. First, what types of pattern appear to be present? For instance, if the graph suggests 7-day cycles in mood, this observation will help the researcher set up the periodogram or spectral analysis (see Chapters 5 and 6) in a way that is best suited to detect this pattern. On the other hand, if there is not much variability in the time series, the researcher might decide at this preliminary stage that a more complicated time-series analysis would be a wasted effort. Second, preliminary examination can help the researcher decide whether there are serious violations of assumptions about data structure (such as normal distribution shape) that might require data transformations or other precautions. Third, preliminary examination of the data can include assessment of trend and other patterns that need to be removed from the time-series data prior to fitting other pattern components such as cycles.

Initial Look at a Graph of Time-Series Data

It is important and useful to do basic data screening and to look at a graph of the time-series data to get a qualitative sense of the patterns it may contain, before doing statistical analyses to describe patterns. For example, consider the data in Figure 3.1 (on p. 42). Numbers of airline passengers (given in thousands) were graphed for each month from the year 1949 to 1960. It is apparent from this graph that there is an increasing trend (numbers of passengers tended to increase consistently from 1949 to 1960). It also looks as if there may be a 12-month cycle, as there are fairly regularly spaced peaks in this graph. These two qualititative features of the graph tell the researcher that at least two kinds of "pattern fitting" will probably be needed to explain tha pattern in the data: a linear trend and a 12-month cycle.

Preliminary data screening should include examination of a histogram and basic descriptive statistics (mean, variance, minimum, maximum, skewness, etc.) for the time-series variable. This examination allows the investigator to judge whether the observed time-series variable has reasonably normally distributed values and to identify any outliers that may require special handling. For example, a time series of reaction time measurements would probably have some relatively extreme values of reaction time; log transformations are fairly effective at reducing these outlier values. Some extreme outliers may be measurement errors that need to be removed or replaced. (Obviously, the more adjustment the researcher has made for outliers, the more caution he or she must use in interpreting any observed patterns.)

The researcher typically hopes that the results from this preliminary screening do not suggest serious problems with the data or violations of assumptions. The time-series variable should be approximately normally distributed, and it should not have any extreme outliers. For physiological time-series data, systematic methods of screening out artifacts may be required in order to identify any data points that appear to be questionable (due to movement artifact, other bioelectric activity, or equipment problems). There should also be a reasonable amount of variability in the scores over time; trends and cycles cannot account for an amount of variance large enough to be of any practical importance unless the original time series itself has a reasonable amount of variability in it. If the analyst is interested in detecting cycles, he or she should look to see whether there appear to be regularly spaced peaks or troughs in this graph that might correspond to the cycles of interest. Finally, the presence of a trend or of heterogeneity of variance across time should be noted so that future analyses can take these potential problems into account. Excellent advice about preliminary data screening is given by Tabachnick and Fidell (1996, Chap. 4).

Following is a summary of basic questions the researcher should ask when examining a plot of a time series.

Are There Extreme Outliers?

Checking for outliers can be a means of identifying measurement errors or artifacts, but it is not sufficient alone. Other data quality control measures should be used as necessary to prevent instrumentation errors. An extreme outlier can have a disproportionate impact on the results of trend analysis and spectral analysis, just as it does in regression or correlation analysis. Outliers that represent measurement errors should be corrected or deleted. Even if the outlier does represent an accurate measurement, it may be advisable to reduce its size, perhaps by rounding the extreme observed value down to the next largest valid data value, or by taking the log of the entire time series.

In later analyses, such as spectral analysis, the researcher may see a large peak in the spectrum that corresponds to a cycle (with a period of 3 minutes, for instance). It is tempting for the researcher to jump to the conclusion that there are cycles 3 minutes long in the time-series data. However, this should not be assumed to be true! The researcher should go back and look at the plot of the time-series data. It is possible that the time-series data do not show a regularly repeated 3-minute cycle. A peak in the spectrum (corresponding to a cycle of 3 minutes) can also be produced by two or three extreme outliers that just happened to be 3 minutes apart. If spectral analysis suggests a cyclic component, it is important to go back to the plot of the time-series data to assess whether there are reasonably regular repetitions of that cycle in the data. For instance, the researcher should count the number of observations between peaks and minima that can be picked out by visually examining the data to see if this matches the cycle length that the spectrum seemed to suggest was a good fit.

Is There Sufficient Variance?

Some time series may be better described as having "not much variance," instead of by fitting trends or cycles. For instance, Warner and Stevens (1991) monitored each subject's systolic blood pressure (SBP) for 40 minutes, obtaining a measure of mean SBP once every 2 seconds. For some subjects, SBP had an extremely wide range (e.g., from 100 to 160 mmHg) and correspondingly large variance. With this amount of variability, it is possible to find trends or cycles that reflect large enough changes in SBP to be of some clinical or practical importance. It might be possible to find cycles with amplitudes as large as 10 or 20 mmHg in a time series that has this much variance. However, for some subjects, SBP varied quite little (e.g., a range of 118 to 123 mmHg). Any "cycles" that might be found in such data would have to have very small amplitudes (on the order of 1 or 2 mmHg), and this is too small to be of much clinical or practical significance. Even if patterns such as trend or cycles can be found in such data, the amount of variability they could account for is limited, and too small to be considered very interesting. It would be better simply to describe this subject's SBP as showing very little variance, rather than attempting to describe patterns that account for very little variance.

Does the Time Series Show Any Apparent Predictable Patterns over Time?

A time series with a large variance can be nonperiodic. However, some time series may show one strong periodic component (as in Figure 1.1) or may show a mixture of two or more periodic components. A linear or curvilinear trend may also occur, either with or without cycles superimposed on it. Notice whether peaks tend to occur at regularly spaced intervals; this is an indication of possible periodicity.

How Is Stationarity of a Time Series Assessed?

A formal mathematical definition of stationarity requires that all parameters of the time series (such as the mean, the variance, the lagged autocorrelations, and so forth) be constant over time. The simplest methods for assessing stationarity of a time series are as follows: First of all, we need to assess whether there is a trend (change in level over time). If a trend is present, it needs to be removed before assessing periodic components. Second, we need to assess whether the variance of the time series is stable (or homogeneous) over time. This can be done given a reasonably long time series ($N > 100$). The time series is divided into several segments of equal length; the variance of the observations within each segment is computed; and a homogeneity of variance test (such as the Levene statistic available in the SPSS one-way ANOVA program)is used to assess whether the variance is homogeneous across segments.

Note that this latter test for homogeneity of variance, like most statistics, assumes independence of observations. Many time-series observations violate this assumption of independence of observations, and therefore the statistical significance estimated for this test may be biased; the test may be either too liberal or too conservative (Kenny & Judd, 1996). However, the test may still be useful as a guide even if the precise risk of Type I error cannot be determined.

If the variance of the time series is not stable across segments, this could be due to one or more extreme outliers that inflate the variance within some segments of the time series. If this is the case, a simple data transformation such as taking the log of the time series may remedy this problem, or it may be appropriate to modify or remove some outliers from the time series.

A more subtle aspect of stationarity is the requirement that the nature of the serial dependence among neighboring observations (e.g., the lagged autocorrelations, or periodicities if any) should be the same across different segments of the time series. A formal test for stationarity in physiological time-series data was suggested by Weber et al. (1992). If the number of observations in the time series is reasonably large ($N >$ 100), this assumption can be checked in a similar manner by computing lag 1 autocorrelations (see the next section of this chapter), or estimates of amount of variance due to particular cycles, separately for each segment of the time series to see if these aspects of pattern are uniform across time (see Warner, 1992a, for an empirical example).

Testing for Significant Sources of Pattern in Time-Series Data

Before carrying out more complex tests for pattern (such as spectral analysis) it may be useful to do a simpler preliminary assessment of whether there is any pattern present in the data at all. The null hypothesis for this test is that the time series consists of white noise: observations uncorrelated with each other. Later when spectral analysis is introduced (in Chapter 6), white noise will be defined more formally as an equal mixture of all the frequencies; that is, no individual periodic component explains a larger share of the variance than the other periodic components.

Lagged Autocorrelation Function

A common tool for assessing pattern in a time series is the lagged autocorrelation function. Most computer packages now provide lagged autocorrelation function procedures. A lag 1 autocorrelation is obtained by correlating the variable X_t (the value of X observed at time t) with a new variable, X_{t-1}, which contains the value of X_t lagged one observation, that is, the value of X seen one observation earlier. If the time-series variable looks like this:

Time (t)	Observed value (X_t)
1	5
2	8
3	4
4	1
5	6

then the lag one autocorrelation is obtained by correlating the variables X_t and X_{t-1} (note that one observation is lost):

t	X_t	X_{t-1}
1	5	—
2	8	5
3	4	8
4	1	4
5	6	1

More generally, we compute a lag k autocorrelation by correlating X_t with X_{t-k}. Usually lagged autocorrelations are reported for lags = 1, 2, 3, . . . up to about $N/4$, where N is the length of the time series. The SPSS autocorrelation procedure (ACF) provides both a table of the lagged autocorrelation for each time lag and a graph of the lagged autocorrelations as a function of the lag k. A 95% confidence interval (CI) can be set up around 0 to assess whether the lagged autocorrelation at each time lag is significantly different from zero. Most computer programs provide this CI.

If the time-series data are "white noise," then most of the lagged autocorrelations should fall within this 95% CI. If many lagged correlations fall outside this CI, then there is some evidence of pattern. Note, however, that even for true white noise, 5% of the lagged autocorrelations would fall outside a 95% CI around 0. If only one or two of the lagged autocorrelations at higher lags are significant, this may not be indicative of a pattern; but if autocorrelations at lags 1 or 2 are statistically significant, then this usually means some pattern is present in the time-series data. If the lagged autocorrelation function oscillates (shows a cyclic pattern), this may suggest that a cycle is present in the original time-series data. If there are large autocorrelations at large time lags, this may be evidence of a trend or other types of nonstationarity in the time series.

Box–Ljung Q Test

Instead of looking at each individual lagged autocorrelation to see if it is significant, the data analyst may want an overall test that examines a set of lagged autocorrelations. The SPSS autocorrelation procedure reports the Box–Ljung Q statistic at each lag. This statistic tests whether the entire set of lagged autocorrelations, ranging from lags of 1, 2, . . . up to lag m, is significantly different from zero. For lag m, the simplest formula for this test statistic is

$$Q = N \sum_{k=1}^{m} r_k^2$$

where N is the number of observations in the time series and m is the number of lagged r's included in the sum (Chatfield, 1991, p. 62).

If we are testing the null hypothesis of white noise for a raw time series, this Q statistic is distributed as χ^2 with m df, where m is the number of lagged autocorrelations included in the set to be tested. If it is significant, then we can conclude that there is some kind of pattern present in our time-series data. This may, however, be due to a number of different kinds of pattern including, but not limited to, linear and curvilinear trends, as well as cycles. Further analysis are required to assess what kind of pattern best describes the time-series data. The Q statistic can be applied to the residuals from model fitting at any stage in the analysis to assess whether the model fitted so far has accounted for all the pattern in the data or whether additional pattern is still present in the residuals.

Durbin–Watson Statistic

Another simple test of a pattern in data that is occasionally used to assess whether residuals from a regression are uncorrelated is the Durbin–Watson statistic. This statistic only assesses correlations between observations separated by a lag of one time unit, so it may not detect more subtle violations of independence. This test statistic has a range from 0 to 4. Values less than 2 indicate a positive lag 1 correlation among the residuals; values greater than 2 indicate negative lag 1 correlation among residuals. The SPSS TRENDS manual includes a table of critical values for the Durbin–Watson statistic. Chatfield (1991, pp. 62–63) provides a brief explanation of this statistic, which is commonly reported by many regression programs as a test of whether residuals are independent.

Trend Description and Trend Removal

Before looking for cycles in time-series data, it is almost always necessary to identify and remove any trends in the time series. One popular method of trend removal involves differencing the time series, a method favored by econometricians. Differencing will be explained only briefly here. In this book, another method of trend description and removal—ordinary least squares (OLS) regression trend analysis—will be used, as differencing can lead to problems in some research situations. This method (OLS regression) is more familiar to behavioral scientists; it involves the use of regression to fit a trend to the time-series data. The observation number t ($t = 1, 2, 3, \ldots, N$) is used as the independent variable; the observed time series X_t is used as the dependent variable. This is the method that is used and recommended for trend analysis and trend removal throughout this book.

Differencing

Differencing has been widely used in econometric time-series analysis, particularly by analysts who are fitting ARIMA (autoregressive integrated moving average) models to time-series data (Box & Jenkins, 1970). Differencing a time series simply involves computing the differences between adjacent values.

If the original time series is

$$1 \quad 3 \quad 5 \quad 7 \quad 9 \quad 11$$

then the first-order differences would be

$$(3–1) \qquad (5–3) \qquad (7–5) \qquad (9–7) \qquad (11–9)$$
$$2 \qquad\quad 2 \qquad\quad 2 \qquad\quad 2$$

Note that in the previous example, the original time series consisted of a linear trend line with a slope of 2. After differencing, the first-order differences no longer have a trend and their mean value of 2 is just the slope of the trend that was present in the original time series.

One nice feature of differencing is that it can be used to remove higher-order polynomial trends and/or cycles. For instance, if the original time-series data consist of a quadratic (X^2) trend, as in this series:

$$1 \quad 4 \quad 9 \quad 16 \quad 25 \quad 36$$

then differencing the series once yields

$$3 \quad 5 \quad 7 \quad 9 \quad 11$$

and differencing the first-order differences again (second-order differencing) yields

$$2 \quad 2 \quad 2 \quad 2$$

Note that if the original time series contains a polynomial trend that corresponds to X raised to the p power, then differencing the time series p times will remove this curvilinear trend.

Furthermore, if the original time series is periodic or cyclic, as in the following example,

$$1 \quad 4 \quad 7 \quad 4 \quad\quad 1 \quad 4 \quad 7 \quad 4 \quad\quad 1 \quad 4 \quad 7 \quad 4$$

then differencing the observations that are one cycle apart, in this case, $X_t - X_{t-4}$, will remove the cycles. This is often used in econometric time-series analysis as a means of removing cycles due to seasonal variation, prior to doing other time series regression analyses.

Advantages of differencing as a tool for trend removal are its simplicity and its flexibility. In addition, differencing is relatively insensitive to the effects of extreme outliers. However, because there are also some disadvantages, OLS trend analysis will be recommended here and used in all subsequent examples. Disadvantages of differencing as a tool for trend removal include the following:

1. It is less convenient to assess the percentage of variance in the trend accounted for in the original time series when trends are removed by differencing than when using OLS trend analysis.

2. In some cases, differencing a time series may "overcorrect" for trend and may introduce artifactual negative correlations between neighboring observations. Differencing should not be used when this occurs.

3. After differencing, the differenced series does not have the same meaning as that of scores in the original time series. The original X_t values were the levels of the measured variable at time t. The $(X_t - X_{t-1})$ differences are *changes* in the level over time. If we use differences in subsequent analyses, we are now making statements about the cyclic behavior of *changes* in X, not in X itself. Under some circumstances this may be what we want to do, but in many situations it may not be a convenient way to describe the data.

4. When we have time-series data for more than one subject, we need to have a data analysis strategy that is consistent across subjects if we want to compare or summarize results across subjects. Because differencing is sometimes an "overcorrection" for trend, we may find that we need to difference the time series for some subjects and not for others. This would mean that for some subjects our subsequent analyses would be examining the behavior of changes in X, whereas for others our subsequent analyses would be examining the behavior of the X scores themselves. These are not directly comparable.

On the other hand, if we use OLS analysis to remove any linear trend from the time-series data for each subject and then do all subsequent analyses on the trend residuals for each subject we can maintain comparable data analysis across subjects. If a subject does not have any trend, then the fitted slope will be close to zero and the trend removal will have little effect on the time series.

Ordinary Least Squares Regression for Trend Removal

The simple bivariate linear regression $X'_t = a + bt$ (where t is the observation number or time index) gives the line that describes any linear trend present in the time series. If there are curvilinear trends, powers of t (t^2, t^3, and so forth) can also be included as predictors. The estimation of a, b, and the multiple correlation square $\underline{R}^2$ (see Table 3.1 in the next section) that is accounted for by trend can be performed using any OLS regression program. Significance testing for this trend model is a bit problematic, however; the residuals from the trend may not be independent of each other, and if they are significantly correlated, the F test using the Mean Square residuals is not valid. It is possible to do simple empirical testing to assess whether the residuals are correlated: either the Box–Ljung Q or the Durbin–Watson statistic, described above, can be used for this purpose.

Empirical Example of Trend Analysis: Airline Passenger Data

Monthly data on numbers of international airline passengers were reported (in 1,000's) for 12 years, or a total of $N = 144$ observations (from Box & Jenkins, 1970). The complete airline passenger data set is given in the form of an SPSS for Windows worksheet (named "airline.sav") in Appendix A. A plot of these raw time-series data is shown in Figure 3.1.

Visual examination of this graph suggests several qualitative de-

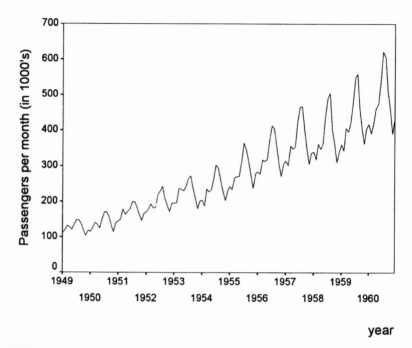

FIGURE 3.1. Graph of raw data on numbers of airline passengers per month (in 1,000's).

scriptions. First of all, there is evidence of a strong linear trend. In addition, 12-month cycles are apparent, although the amplitude of these cycles is rather small compared to the magnitude of the trend. It also appears that the variance of the time series is increasing over time (the amplitude, or distances of peaks and troughs of the cycle from the trend line, tends to increase over time). Subsequent chapters will describe quantitative methods that enable us to give a more precise description of these features of the data. An initial check can be done using the lagged autocorrelation function to assess whether there is a significant "pattern" in this time series.

Because there is clearly a strong linear trend in this time series, the trend component was removed from the time series before doing further analysis to look for other patterns such as cycles. If this were not done, the trend component would be the dominant feature seen in the lagged autocorrelation function. In Chapter 11 we will see that a trend component produces a broad peak at the low-frequency end of the spectrum that makes it difficult to detect any cycles that might be present.

OLS trend removal was used to fit and remove a linear trend from the airline passenger time-series data. Visual examination of the graph in Figure 3.1 suggests a strong positive linear trend, with numbers of passengers increasing over time. The trend does not appear to be curvilinear; a linear model will probably give an adequate fit. The 12-month cycle in amount of passengers is also evident in this plot, although the amplitude of the 12-month cycles is rather small compared to the trend. Before doing an analysis to assess this 12-month cycle, we need to describe and remove the trend.

The observation number (ranging from 1 to 144) was included as a variable in the SPSS worksheet. The SPSS output file (in Table 3.1) shows the results when a regression analysis is done to predict the raw time-series data X_t (numbers of airline passengers per month, in 1,000's) from the observation number t. The regression results showed that linear trend accounted for 85% of the variance in the time series. The equation that describes this trend in passengers is $X_t' = 90.302 + 2.66 \cdot t$. Figure 3.2 shows a graph of the raw time-series data with the fitted linear trend component superimposed on it.

The residuals from this trend analysis were saved as a new variable in the SPSS worksheet (the variable containing these residuals from linear trend was named "tres"). Next, lagged autocorrelation analyses were

TABLE 3.1. Results of Linear Trend Fitting Using SPSS Regression (for Airline Passenger Data)

```
Multiple R            .92397
R Square              .85371
Adjusted R Square     .85268
Standard Error      46.04655
```

Analysis of Variance

	df	Sum of Squares	Mean Square
Regression	1	1757088.11229	1757088.11229
Residual	142	301080.44327	2120.28481

F =   828.70381   Signif F = .0000

---------------- Variables in the Equation ----------------

Variable	B	SE B	Beta	T	Sig T
OBS	2.657383	.092311	.923967	28.787	.0000
(Constant)	90.302682	7.634627		11.828	.0000

Note. In this book, $\underline{R}$ (with underline) refers to the multiple correlation or goodness of fit measure in a regression; R (no underline) refers to the amplitude of a sinusoid. OBS is the variable that represents "observation number" (OBS = 1, 2, 3, . . . , 144).

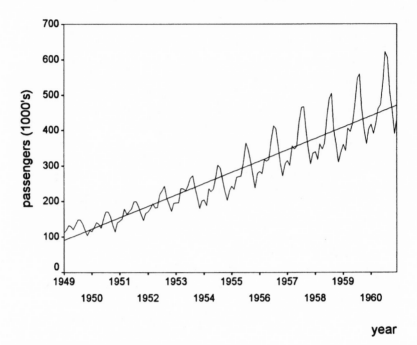

FIGURE 3.2. Graph of time-series data on airline passengers with the OLS linear trend component superimposed on it.

run on these residuals from this trend analysis, using SPSS TRENDS program commands to select (graph) (time series) (autocorrelation function) from the menu.

The resulting lagged autocorrelation function (ACF) for the trend residuals of the airline passenger data were graphed in Figure 3.3. The accompanying SPSS output file includes a table of the lagged autocorrelations and the Box–Ljung Q test (this output file is included in Table 3.2).

There are significant correlations among residuals at various time lags. In particular, the cyclical form of the ACF and the presence of a large and statistically significant autocorrelation at a lag of 12 months suggest the existence of a 12-month cycle.

Because many of the lagged autocorrelations are statistically significant and the Box–Ljung Q test was also significant at all lags ranging from 1 to 16, the assumption of independence among residuals is clearly violated. This has two implications: First, there is an additional pattern in the data, beyond the linear trend component that has already been identified. Subsequent analyses, described in later chapters, can confirm

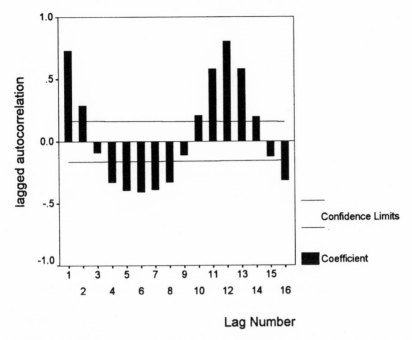

FIGURE 3.3. Lagged autocorrelation function (ACF) for residuals from linear trend in the airline passenger data. A 12-month cycle is evident.

that this pattern is primarily due to a 12-month cycle. Second, we cannot use an ordinary F ratio to assess the significance of our fitted trend. Because significance testing is not the main goal at this stage, we will not go into a lengthy digression about the kinds of additional model fitting one would need to do to obtain the "white noise" residuals that could be used to test trend significance; instead, in this empirical example, the $\underline{R}^2$ for trend was reported as descriptive information without assessing its statistical significance.

Additional follow-up analyses are illustrated in Figure 3.4 and Table 3.3. These are simple analyses that can be done without any knowledge of more advanced techniques in spectral analysis. The following questions arise: What is the "shape" of the 12-month cycle? Specifically, during what months are passenger numbers high or low? To assess this, the passenger totals for each month (after removal of the linear trend) are averaged across years. For example, the value graphed on the Y axis for month 1 (January) in Figure 3.4 is simply the mean frequency across all 12 Januaries (after removal of the linear trend). From this graph it is easy to see that the months that typically have the highest amounts of pas-

TABLE 3.2. Lagged Autocorrelations for Trend Residuals (for Airline Passenger Data)

Lag	Auto- Corr.	Stand. Err.	-1 -.75 -.5 -.25 0 .25 .5 .75 1	Box-Ljung	Prob.
1	.728	.082	. |**.************	77.989	.000
2	.288	.082	. |**.***	90.282	.000
3	-.089	.082	.**| .	91.451	.000
4	-.329	.082	****.**| .	107.657	.000
5	-.393	.081	*****.**| .	131.043	.000
6	-.406	.081	*****.**| .	156.115	.000
7	-.388	.081	*****.**| .	179.234	.000
8	-.328	.080	****.**| .	195.916	.000
9	-.110	.080	.**| .	197.817	.000
10	.207	.080	. |**.*	204.567	.000
11	.580	.080	. |**.*********	257.686	.000
12	.802	.079	. |**.**************	360.230	.000
13	.579	.079	. |**.*********	414.051	.000
14	.197	.079	. |**.*	420.330	.000
15	-.123	.078	.**| .	422.807	.000
16	-.316	.078	***.**| .	439.174	.000

Plot Symbols: Autocorrelations *   Two Standard Error Limits .

Note. These are shown as a graph in Figure 3.3; exact values are reported here.

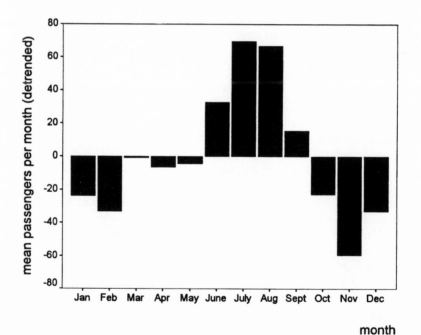

FIGURE 3.4. Mean number of passengers for each month (in 1,000's) after removal of linear trend component from time series. Once the trend component has been removed, a 12-month cycle with highest passenger levels in summer is evident.

senger traffic are July and August whereas the winter months typically had relatively low amounts of passengers.

Another question has to do with the stationarity of the time series. We have already seen that there was a trend that had to be removed, but another requirement of stationarity is homogeneity of variance over time. A simple way to assess this is to do an ANOVA using the years as the levels of the independent variable. Each group in the ANOVA in Table 3.3 is the data for 1 year; each group contains the passenger frequencies for the 12 months in that year. The Levene test for homogeneity of variance, available in SPSS as a feature in the ANOVA program, indicates that there is a significant difference in the variances for these 12 years; in fact, the variance tends to increase over time. A visual examination of the graph of the original time-series data in Figure 3.1 indicates that in addition to an increase in the mean over time, there is also an increase in the *variance* in passenger load over time. The amplitude of the cycles appears to be increasing over time. This is a violation of the assumption of homogeneity of variance (which is one

TABLE 3.3. One-Way ANOVA Comparing Mean Numbers of Airline Passengers across Years (after Removal of Linear Trend from Time Series)

Source	df	Sum of Squares	Mean Squares	F Ratio	F Prob.
Between Groups	11	23754.1121	2159.4647	1.0278	.4256
Within Groups	132	277326.3312	2100.9571		
Total	143	301080.4433			

Group	Count	Mean
Grp1949	12	21.7484
Grp1950	12	2.8598
Grp1951	12	1.4712
Grp1952	12	-3.5841
Grp1953	12	-7.4727
Grp1954	12	-25.4446
Grp1955	12	-12.2499
Grp1956	12	.1115
Grp1957	12	8.3896
Grp1958	12	-10.9157
Grp1959	12	4.6124
Grp1960	12	20.4738
Total	144	.0000

Levene Test for Homogeneity of Variances

Statistic	df1	df2	2-tail Sig.
4.6792	11	132	.000

of the characteristics a time series must have to be considered stationary).

The periodogram and spectral analysis described in Chapters 5 and 6, respectively, will describe the *average* amplitude of the cycle; these analyses by themselves obscure the fact that the amplitude is changing over time. A log transformation of the data might somewhat reduce this heterogeneity of variance. However, one suggested way to deal with this difficulty is to include a discussion of the increase in variance over time in the description of the data—and to be aware that the amplitude (R) that is estimated by the periodogram analysis in Chapter 5 would be an "average" amplitude that is in fact changing over time. More advanced techniques that are briefly introduced later (Chapter 7), such as complex demodulation, will make it possible to describe the change in amplitude across time more precisely.

Chapter Summary

To summarize, this chapter has described and demonstrated methods for identification and removal of trend in time series, and a method for assessment of homogeneity of variance. Not every time series has a trend component that has to be removed before performing periodogram analysis or spectral analysis; but many do have such components. In subsequent chapters, it is generally assumed that the analyses that describe cycles (periodogram analysis and spectral analysis) are being applied to residuals from the trend—except in a few cases where the original data did not have any trend that needed to be removed. If the trend is not removed, it produces artifactual patterns in a periodogram or spectrum that make it difficult to see any cycles that might be present, so trend removal is an essential step in time-series data analysis. Many econometricians prefer to use differencing as a means of trend removal; however, use of OLS regression is recommended here as the means of trend removal.

Subsequent chapters describe how cycles can be detected and described in time-series data (after trend has been removed). Chapter 4 introduces harmonic analysis: a method of estimating the parameters of a sinusoidal cycle, if the period is already known a priori. Chapters 5 and 6 go on to describe periodogram and spectral analysis, which are exploratory methods used to assess what cycle length (or lengths) account for most of the variance in a time series. These techniques are more useful when the cycle length is not known a priori, which is often the situation in behavioral time-series research.

CHAPTER 4

Harmonic Analysis

Introduction

After preliminary examination of the data and a review of past time-series research on the particular variable that is of interest, the researcher may be in one of two situations. One possibility is that the cycle length that should fit the time-series data is known. This knowledge may be available because the cycles are so regular that the cycle length can be seen by looking at a graph of the time series (e.g., peaks and troughs in the sunspot data occur at 11-year intervals in Figure 1.1, suggesting that an 11-year cycle is the best fit to this time series). Alternatively, knowledge of the best-fitting cycle length may be available from past research (as in the research on 7-day cycles in mood or approximately 24-hour cycles in many physiological processes).

If the cycle length is known, then the researcher can use harmonic analysis to model the cyclic component of the time series. Just as trend analysis involves estimating a slope and intercept to obtain the trend line that best fits the time-series data, harmonic analysis involves estimating the amplitude and phase of a cycle (of a particular known period) that best fits the time-series data. Generally, any trend that was identified by earlier exploratory analyses (as described in Chapter 3) is removed from the time series; modeling of periodic components is applied to residuals from trends. When the cycle length or period is known, harmonic analysis provides estimates of the other three parameters (mean, phase, amplitude of the sinusoid) that maximize fit to the observed time series. When the cycle length of a time series is known a priori, harmonic analysis is an appropriate analysis.

When the cycle length is not known a priori—either because there is little or no past research or because the cycles are not regular enough to be identified easily from visual examination of a graph of the time series—then exploratory methods such as periodogram analysis (Chapter 5) and spectral analysis (Chapter 6) may be a more appropriate place to begin. These procedures allow the researcher to fit an entire set of cycles of different lengths or periods, as a way of assessing whether any cycle length provides a good fit to the observed time series.

However, the reader is urged not to skip this chapter on harmonic analysis even if he or she does not know the cycle length a priori. The information in this chapter is useful and important even in research situations where the cycle length is not initially known. Harmonic analysis is presented before periodogram and spectral analysis in this book because it is a simpler method, and it is, in effect, the building block from which the more complex analyses in Chapters 5 and 6 are constructed. Periodogram analysis can be understood as a set of several harmonic analyses. In addition, harmonic analysis can be a very useful follow-up analysis after a periodogram or spectral analysis has been used to identify cyclic components. A periodogram analysis or spectral analysis can identify one (or several) approximate cycle lengths that account for relatively large proportions of the variance in the time-series data. The estimated cycle length obtained from a periodogram analysis can then be used to do a harmonic analysis. The estimate of the period can be improved by using grid search methods: The researcher can vary the estimate of the cycle length or period and try doing a harmonic analysis to fit cycles with different periods, until a cycle length or period is identified for which the goodness of fit is particularly strong.

If the cycle length of the time series is known a priori, then a harmonic analysis of the time series is an appropriate next step in data analysis after trend removel. The cycle length is assumed to have a specific value, and the other three parameters of the sinusoid (mean, amplitude, and phase) are estimated using least squares methods. However, if the researcher believes that the data may be cyclic but does not know what cycle lengths to expect, then techniques covered in later chapters—such as periodogram analysis (Chapter 5) and spectral analysis (Chapter 6)—are appropriate next steps in analysis because they can be used for identification of the (unknown) lengths or periods of cycles.

Harmonic Analysis as a Type of Regression Analysis

Harmonic analysis can be most easily understood by comparing it to more familiar linear regression. Suppose that we want to model a linear trend in a time series, for instance, daily mood ratings of a single subject obtained once a day for N days. The variable t stands for the observation number or day $(1, 2, 3, \ldots, N)$. The variable X_t stands for the mood rating obtained on day t. We can set up a simple linear regression equation to predict X_t from t:

$$X'_t = a + bt$$

This is a simple bivariate linear regression equation, and the ordinary least squares (OLS) regression procedures learned in elementary statistics courses can be applied to this problem. We can obtain an intercept and slope, which are the parameters for this fitted line; we can obtain an $\underline{R}^2$ (squared multiple) to assess the percentage of variance in the X_t time series that is due to linear trend.

To assess whether the X_t time series tends to show regular, sinusoidal cycles, we fit a somewhat more complex model, replacing t with a trigonometric function of t that represents a sinusoid and adding the parameters that allow us to estimate the phase and amplitude that best fit the observed time series. This model is the basis for harmonic analysis. The general equation for a harmonic analysis that involves fitting a sinusoid to a time series is

$$X_t = \mu + R \cos(\omega t + \phi) + \epsilon_t$$

where X_t is the observed value of X at time t; μ is the mean or level of the time series; R is the amplitude or height of the waveform; ϕ is the phase, or location of peaks relative to time zero; ϵ_t are residuals that are unrelated to the fitted cycles; and t is the observation number (0, 1, 2, ..., N).

The ω term in the foregoing equation ($\omega = 2\pi/\tau$) requires a bit more explanation. The factor of 2π included in the cosine function is needed to convert the frequency $1/\tau$ into radians. Arguments for trignometric functions can be expressed in terms of degrees as well as in radians; but in spectral analysis it is generally assumed that the angles are expressed in radians.

The term τ is the period or cycle length of the cosine function that is being fitted. A cosine function that has a cycle length of τ has a frequency of $1/\tau$ For instance, if the researcher has a time series where each observation is mood measured once a day, and if the researcher wants to fit a 7-day cycle, then the value of τ is set at 7; the values of the cosine function can then be generated for various values of the time variable t. The remaining parameters are then estimated using OLS regression methods. A complete and more technical treatment of the estimation procedures is given by Bloomfield (1976, Chap. 2). The basic equations are reproduced on the next few pages, after the introduction of some notation.

There are two ways of writing the equation for a sinusoid. The version presented earlier was as follows:

$$X_t = \mu + R \cos(\omega t + \phi) + \epsilon_t$$

An alternative way of writing this equation is more convenient for computation of parameter estimates. Another way of representing the X_t time series as a function of a sinusoidal waveform with period τ, amplitude R, and phase ϕ is an equation that includes both a sine and a cosine term, as follows:

$$X_t = \mu + A \cos(\omega t) + B \sin(\omega t) + \epsilon_t$$

where μ is the mean or level of the X time series; τ is the predetermined period or cycle length (given as number of observations per cycle); ω is the corresponding frequency in radians ($2\pi/\tau$; $\cos(\omega t)$ is the cosine function of period τ evaluated at all values of t; $\sin(\omega t)$ is the sine function of period τ evaluated at all values of t; ϵ_t are residuals, uncorrelated with the cos and sin terms; and t is the time counter or observation number $(0, 1, 2, \ldots, N)$. This equation is just an alternative way of representing the most general case of the sinusoid. By varying the parameters A and B that indicate how much weight to give to the sine and cosine components, it is possible to generate a sinusoidal waveform (of period τ) with any particular amplitude and phase. Sine and cosine functions have the same waveform shape, but they differ in phase (the cosine has a peak value of 1 at $t = 0$, whereas the sine has a peak value of 1 a quarter of a cycle later). The sine and cosine function of period τ are orthogonal to each other (that is, they are uncorrelated). Therefore, by varying the relative size of the A and B coefficients it is possible to model a sinusoidal waveform with any phase and amplitude. Together the sine and cosine functions form a basis (in linear algebra terms) for the set of all possible sinusoids. By varying the A and B coefficients of this equation, it is possible to generate all possible sinusoids of period τ—sinusoids with any mean μ, amplitude R, and phase ϕ. Given an assumed value of τ, the cycle length or period, the remaining three parameters (the mean, amplitude, and phase of that cycle) can be estimated using OLS regression methods (as shown in Bloomfield, 1976, Chap. 2) to obtain the best possible fit to an observed time series.

The overall amplitude (R) of the waveform represented by this equation is a function of the magnitude of the A and B coefficients:

$$R = (A^2 + B^2)^{1/2}$$

The phase (ϕ) of the waveform represented by this equation depends on the relative sizes of A and B. The formal equation for ϕ is $\tan(\phi) = -B/A$. The function $\arctan(-B, A)$ equals ϕ (Bloomfield, 1976, pp. 12–13).

If we arbitrarily assume a particular value of τ, for instance, if we decide that we want to fit a 7-day cycle to our daily mood time-series data, then the remaining parameters of this model (the mean μ and the coefficients A and B) can all be estimated using OLS regression methods (Bloomfield, 1976, pp. 11–14). Then the overall amplitude R and phase ϕ can be estimated from the A and B coefficients that are obtained when the foregoing equation is fitted to the time-series data using OLS regression.

Essentially, harmonic analysis involves the creation of two predictor variables: one time series that represents a sine of period τ, and another that represents a cosine of period τ. For now, we are assuming that τ, the cycle length or period, is known a priori. This is often the case in research on weekly, daily, or monthly cycles. Later chapters will address the question of how to assess which of many possible periods are good fits to the time time-series data when the best value of τ is not known a priori.

A sine and cosine function of the same period are orthogonal to each other; because they are independent or uncorrelated, when they are assessed as regression predictors there is no need to control for any redundancy between these predictors. We can assess the goodness of fit of the sine and cosine function each independently by calculating the sum of cross products between the sine template and the observed X_t time series, and the sum of cross products between the cosine template and the observed X_t time series. From Bloomfield (1976, p. 14) the estimates of the parameters μ, A, and B are found as follows:

$$\tilde{\mu} = \frac{1}{N} \sum_{t=1}^{n} X_t$$

$$\tilde{A} = \frac{2}{N} \sum_{t=1}^{n} (X_t - \overline{X}) \cos \omega t$$

$$\tilde{B} = \frac{2}{N} \sum_{t=1}^{n} (X_t - \overline{X}) \sin \omega t$$

Notice that to obtain an estimate for the A coefficient, you cross-multiply the deviations from the mean of the observed X time series times the values of a cosine function with period τ (Recall that $\omega = 2\pi/\tau$;

that is, it is the frequency that corresponds to a cycle length of τ, expressed in radians; and frequency is inversely related to cycle length or period.) Essentially, this assesses how strongly the X_t time series tends to covary with the cosine function that we are trying to fit; a large estimated A coefficient implies a good fit between this cosine function and the time series. This is essentially similar to the sum of cross products term that is calculated in a bivariate regression. In fact, these calculations can be done easily using an OLS regression program.

To assess how well a cycle of period 7 days fits the observed time-series data (i.e., to fit a cycle with $\tau = 7$), one could do the following: Generate two predictor variables to represent the sine and cosine functions that correspond to a period of 7 days. Let t be the time counter variable that indicates the day number ($t = 0, 1, 2, \ldots, N$; usually when we want to do computations involving phase it is more convenient to start the time counter at 0 rather than 1). Then the two new computed variables could be obtained by setting up "compute" or data transformation commands to create new time-series variables s_t and c_t, as follows:

$$t = \text{observation number } (t = 0, 1, 2, \ldots, N)$$
$$c_t = \cos(2\pi/7 \cdot t)$$
$$s_t = \sin(2\pi/7 \cdot t)$$

Then to obtain estimates of the weights for the cosine and sine terms, all that it is necessary is to perform an OLS regression analysis to predict X_t from c_t and s_t. The intercept obtained for this regression equation is an estimate of the mean of the time series. The coefficients of c_t and s_t provide estimates of A and B, respectively. The regression statistics, such as the squared multiple $\underline{R}$, would provide an indication of the goodness of fit of the model; the best fitting amplitude and phase can be calculated from the A and B coefficients.

It is unfortunate that the letter "R" is used in the regression literature to denote a multiple correlation and in the harmonic analysis literature to denote the amplitude of a sinusoid. Be careful not to confuse the two different uses of the letter: sometimes $\underline{R}$ refers to the multiple correlation or goodness of fit measure in a regression; at other times, R refers to the amplitude of a sinusoid. Note that henceforth, in this book, $\underline{R}$ (underlined) will represent the multiple correlation; R (not underlined) will represent the amplitude.

Parameter estimation becomes somewhat more complicated (1) if the time series does not contain an integer number of the cycles or (2) if there are two or more cyclic components in the time series. (For instance, when the time series contains a fractional part of a cycle, then the observed mean of the time series is not a good estimate of the overall mean of the time series.) In either of these situations, the estimates of any one parameter (such as the mean of the time series) should be done in a way that takes the values of the other parameters into account. There is no simple analytic solution to this problem, but numerical methods can be used; Bloomfield (1976, p. 23) describes a method he calls "cyclical descent." In effect, the best estimate of each parameter is found using maximum likelihood/grid search. According to Bloomfield, the model parameters can be divided into sets

> in such a way that the optimization with respect to parameters in any one subset, holding the remaining parameters constant, can be done fairly easily. The method then is to update the subsets successively by solving these manageable optimization problems in turn. The basic method cycles among these subsets in some fixed order, until a complete cycle results in an effectively zero change in the function to be optimized. (p. 22)

Bloomfield provided FORTRAN programs that carry out this numerical estimation; this would not be easy to do using a "packaged" program such as SPSS.

In the example below, the length of the time series ($N = 14$) is exactly two times the period (7). This was done intentionally to make it simple to estimate the mean without having to correct for a fractional part of a cycle. In general, estimation of parameters is much easier if the total N (the number of observations in the time series) is an integer multiple of the cycle length.

Significance Testing in Harmonic Analysis

In principle, we could look at the t or F tests to assess the significance of the A and B coefficients in the regression analysis model. More often the researcher is interested in the overall significance of the model (the overall $\underline{R}^2$); and in later chapters it is this overall amount of variance accounted for by each periodic component (rather than the significance of the sine and cosine components) that is of interest. However, it is possible that the residuals for this harmonic analysis model will not be independent or white noise, and in such cases, significance testing raises a problem. Conventional significance testing procedures used with regression analysis assume that the residuals that are used to compute an error variance are independent of each other. Computation of conventional F

or t ratios yields biased results if this independence assumption is violated, and often this leads to inflated risk of Type I error (Gottman, 1981b). A conventional F ratio can be used to assess the significance of the harmonic analysis or multiple regression model only if the residuals from this model are "white noise," that is, independent of each other. A test (such as the Box–Ljung Q statistic, described in Chapter 3) can be used to evaluate whether the residuals are white noise. If the Box–Ljung statistic is not significant, then the conventional F ratio for the overall regression equation can be used to test the significance of the $\underline{R}^2$.

Empirical Example: Harmonic Analysis of Mood Data

A complete example of harmonic analysis, using the simulated mood data introduced in Appendix 2.1 (Chapter 2), is presented here. Preliminary examination of the graph of these time-series data suggested that a cycle with a period of 7 days might be a good fit to this mood time series. An OLS regression analysis to predict mood from the observation number indicated that there was no linear trend in the simulated mood data; therefore the harmonic analysis in this example is based upon the raw mood scores. In many real-life applications, and in later examples in this book, harmonic or spectral analysis will be performed on residuals from the trend rather than the raw time-series data. Alternatively, the trend component can be included in the complete statement of the model for the prediction of the time series.

Simulated data are used at this point so that the patterns are clear and so there are no violations of assumptions. Later examples will use real data that are more complex and that illustrate common violations of assumptions, thus showing what modifications can be made to the analysis when we try to deal with these problems.

Table 4.1 shows an SPSS for Windows worksheet that contains the day number in column 1 (ranging from 0 to 13) and the (simulated) mood-rating data in column 2. The cosine and sine functions (with periods of 7 days) are shown in the next two columns of the SPSS data worksheet. Transform statements were used to compute the values of these two "template" sinusoid variables that represent a 7-day cycle, with the "day" variable as the time counter. The variable cos 7 was created by computing $\cos(2\pi\text{day}/7)$ for each value of the day variable. The variable sin 7 was created by computing $\sin(2\pi\text{day}/7)$.

To do a harmonic analysis, the SPSS regression procedure was used with the cos 7 and sin 7 variables as predictors of the mood variable. Results of the multiple regression are shown in Table 4.2. The conclusions

TABLE 4.1. SPSS for Windows Worksheet Showing Mood Data and the 7-Day sine and cosine Functions

	day	mood	sin7	cos7
I	0	4.99	.00	1.00
2	I	3.37	.78	.62
3	2	3.06	.97	−.22
4	3	2.14	.43	−.90
5	4	2.91	−.43	−.90
6	5	3.96	−.97	−.22
7	6	4.88	−.78	.62
8	7	5.35	.00	1.00
9	8	2.37	.78	.62
10	9	2.59	.97	−.22
11	10	2.75	.43	−.90
12	11	2.42	−.43	−.90
13	12	3.51	−.97	−.22
14	13	4.70	−.78	.62

from this harmonic analysis would be as follows: The equation that describes the cyclic pattern in mood is

$$\text{predicted mood, day } t = \mu + A \cos(\omega t) + B \sin(\omega t)$$
$$= 3.5 + 1.08 \cos(\omega t) - .66 \sin(\omega t)$$

with the value of ω set at $2\pi/7$. The overall mean of the time series is estimated by the intercept for this equation: 3.5. A 7-day cycle accounts for 76.8% of the variance in the observed time series (this is the multiple R^2 for the regression equation, i.e., the percent of variance in the time series that is accounted for by the sinusoid).

The estimated amplitude (R) of this cycle is calculated by taking $(A^2 + B^2)^{1/2} = 1.27$. This amplitude estimate indicates that the mood cycles have peaks and troughs about 1.27 points above and below the overall mean mood of 3.5.

TABLE 4.2. SPSS Regression Results, Predicting Mood Time-Series Observed Values from the 7-Day sine and cosine Functions

```
* * * * MULTIPLE REGRESSION * * * *

Listwise Deletion of Missing Data
Equation Number 1    Dependent Variable..    MOOD
Block Number  1.  Method:  Enter      SIN7      COS7

Variable(s) entered on Step Number
    1..    COS7
    2..    SIN7

Multiple R           .87640
R Square             .76808
Adjusted R Square    .72591
Standard Error       .55489

Analysis of Variance
                    df        Sum of Squares      Mean Square
Regression           2           11.21685           5.60843
Residual            11            3.38695            .30790

F = 18.21484       Signif F = .0003

---------------- Variables in the Equation ----------------

Variable         B          SE B        Beta         T        Sig T

SIN7         -.655289     .209729    -.453679    -3.124      .0097
COS7         1.083051     .209729     .749834     5.164      .0003
(Constant)   3.499997     .148301                23.601      .0000

End Block Number  1  All requested variables entered.
```

In order to see what features of the time-series data were captured by this harmonic analysis, values for the predicted moods were generated by substituting the day variable ($t = 0, 1, 2, \ldots, 13$) into this regression equation. The predicted moods that are based on this 7-day sinusoidal cycle were thus computed from this equation:

$$3.5 + 1.08 \cos(2\pi/7 \cdot \mathrm{day}) - .66 \sin(2\pi/7 \cdot \mathrm{day})$$

These predicted values of mood were placed in the fifth column of the SPSS worksheet using the variable name "cycle." Figure 4.1 shows the

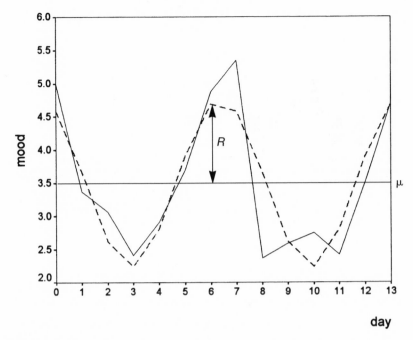

FIGURE 4.1. Mood time series (solid line) with a 7-day harmonic cycle (dashed line) superimposed.

fitted sinusoid (the dashed line) superimposed on the raw data (the solid line). The mean (μ = 3.5) and the amplitude (R = 1.266) are marked on this graph.

The residuals from this model were calculated by creating a new variable (called "res"), which is the raw mood data minus the fitted cycle. This variable was placed in the sixth column of the SPSS worksheet. If the 7-day cycle accounts for all the pattern in these time-series data, then this residual variable should not show any serial dependence or patterning. To assess whether the residuals from this model are in fact white noise, the lagged autocorrelations for the residuals were obtained. These are shown in Figure 4.2.

The lagged autocorrelations and the Box–Ljung Q statistic (shown in the SPSS output file in Table 4.3) were not statistically significant at any lag, so the residuals were essentially white noise. Therefore it is possible to use an F ratio to assess the significance of the harmonic analysis results obtained earlier using the SPSS regression procedure. The F ratio (shown in the SPSS regression output in Table 4.2) was $F(2, 11)$ = 18.28, p = .0003; therefore the results are statistically significant. We can

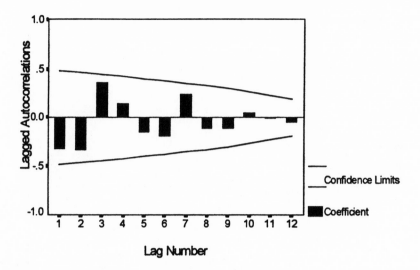

Lag Number

FIGURE 4.2. Lagged autocorrelations for the residuals from a 7-day cycle in the mood time-series data. Because no lagged autocorrelations were statistically significant, the residuals were judged to be white noise.

TABLE 4.3. Lagged ACF for the Residuals from the 7-Day Mood Cycles

```
Autocorrelations:      RES

         Auto- Stand.
Lag   Corr.  Err.  -1  -.75  -.5  -.25    0   .25   .5   .75    1    Box-Ljung  Prob.
                    +——+——+——+——+——+——+——+
  1  -.322  .241              .  ******|                          1.784    .182
  2  -.332  .231            . *******|             .              3.839    .147
  3   .364  .222                .    |******* .                   6.539    .088
  4   .146  .211                  .  |***    .                    7.015    .135
  5  -.154  .200                . ***|       .                    7.603    .179
  6  -.193  .189                .****|       .                    8.644    .195
  7   .239  .177                    .|*****  .                   10.465    .164
  8  -.118  .164                  . **|     .                    10.988    .202
  9  -.117  .149                  . **|   .                      11.605    .237
 10   .045  .134                     .|*  .                      11.719    .304
 11  -.012  .116                     .  *    .                   11.729    .384
 12  -.048  .094                     . *|  .                     11.983    .447

Plot Symbols:      Autocorrelations *      Two Standard Error Limits .

Total cases:  14      Computable first lags:  13
```

conclude that a 7-day cycle is a statistically significant component of the mood time series, and the residuals left over after removing this 7-day cycle are not significantly different from white noise, so no other pattern seems to be present in this mood time series.

Note that, if the phase were changed in this example, the relative sizes of the A and B coefficients which assess the relative importance of sinusoids with the phase of a sine wave versus the phase of a cosine wave would change but the overall value of $A^2 + B^2$ would not change.

Beyond Harmonic Analysis: Preview of Related Methods

This example is artificially simple; real data tend to be much more complex. Subsequent chapters will show how to deal with the additional complexity that is often found in real time-series data. Issues that will be considered include the following:

1. If there are linear and/or curvilinear trends in the data, they may need to be accounted for and removed before we try to assess the presence of cycles. This can be done by differencing (as in econometric time-series analysis), but the approach recommended here is OLS trend analysis and removal.

2. If the period τ is not known a priori, then it may be identified through a more exploratory analysis. Periodogram analysis and spectral analysis provide means of identifying which periods (if any) account for large proportions of variance in the observed time series. Statistical significance tests may also be performed to assess whether a cycle explains a larger amount of variance than would be expected by chance if the data were white noise, or completely random.

3. If there are two or more cyclic components, a more complex description of the data may be needed. Using periodogram analysis or spectral analysis, we can estimate the percent of variance explained by one or more frequency bands and use this as a way of characterizing the patterning of the time series.

4. If there are extreme outliers, data transformations may be required to avoid having results that are disproportionately influenced by a small number of extreme observations.

5. If the data are not "stationary," then some characteristics of the time series are changing over time (e.g., the mean, the variance, and the period or amplitude of any cyclic components) If data are not stationary, then it becomes much more difficult to describe the pattern over the en-

tire time series and it may be quite misleading to attempt to fit one model to the entire time series. Other methods (such as complex demodulation) that can describe these changes over time will be briefly introduced later (Chapter 7).

Chapter Summary

Harmonic analysis provides a method of fitting a cyclic component to a time series; familiar OLS regression programs can be used to carry out this analysis. If there are two or more cyclic components in a time series, each one can be represented as an additional pair of sine and cosine terms in a more extensive model for multiple cycles (see Bloomfield, 1976). The fitted sinusoid(s) can be plotted on the same graph as the raw time-series data, and this provides a clear illustration of the features of the data that are being modeled. Harmonic analysis can be undertaken as a first step in time-series analysis (if the cycle length is known a priori), or it can be done as a follow-up analysis (cycle length estimates can be obtained from the periodogram, Chapter 5, or spectral analysis, Chapter 6). If the initial guess at the value of the period or cycle length is only an approximation, "grid search" or maximum likelihood estimation procedures can be used to try to get a more precise estimate. That is, the researcher can try out various candidate values of τ to see which one provides the best fit to the time-series data; τ does not have to be limited to integer values, and in fact a noninteger value may provide the best fit.

Harmonic analysis will only provide a good "fit" if the time series is stationary. When a time series is stationary, that means (among other things) that the cycle length itself is not changing over time. If the cycle length does seem to be changing over time, one relatively simple method of dealing with this is to conduct harmonic analysis separately on different sections of the time series (assuming that there are enough observations to enable the analyst to look at shorter segments of the data). Complex demodulation (as described by Bloomfield, 1976, Chap. 6) involves plotting the (changing) cycle length or period as a function of time; this may be more appropriate for data sets where cycle length and/or amplitude are changing over time.

The next chapter introduces periodogram analysis: it can be seen as a *set* of harmonic analyses. A set of "Fourier frequencies" is fitted to the time series; the choice of these frequencies is determined by the length of the time series (the number of observations, N) and the sampling frequency (Δt). For each frequency or period, estimates of the corresponding A and B coefficients, as well as an overall measure of goodness of fit, are obtained. Thus, periodogram analysis is just a generalization of harmonic analysis. Instead of estimating parameters for just one sinusoid (as in this chapter), parameters are estimated for a set of many sinusoids, as a means of assessing which, if any, of them explain a large proportion of variance in the time series.

CHAPTER 5

Periodogram Analysis

Introduction

Periodogram analysis is, essentially, a generalization of the harmonic analysis methods presented in Chapter 4. In that chapter, harmonic analysis was presented as a method of regression analysis in which the best-fitting amplitude, mean, and phase of a cycle could be estimated when the cycle length or period was already known. However, in many research situations, the periods of cycles are not known a priori. The researcher's goal is to identify which, if any, periodic components explain a large enough percentage of the variance in the time series to be of interest, and to describe their cycle lengths and amplitudes.

The basic tool for partitioning the variance of a time series (of length N) into variance accounted for by $N/2$ periodic components is a form of analysis of variance (ANOVA) called the periodogram (Box & Jenkins, 1970, pp. 36–39; Kaplan, 1983). In effect, periodogram analysis estimates the percentage of the variance in the time series that is accounted for by each of a set of different sinusoids. In addition, the periodogram provides estimates of the phase and amplitude for each sinusoid.

Generally, any trends in the time series are removed prior to doing periodogram analysis. The analyses described here are generally applied to the *residuals* from the trend analysis. If there is a linear or curvilinear trend componnent in the time series, it will influence the partitioning of variance in the periodogram such that longer cycles (or lower frequencies) will spuriously appear to account for large shares of the variance in the time series (see Chapter 11 for further discussion and an example). In order to detect cycles in a way that gets rid of this source of artifact, it is necessary to remove trends from the data before doing periodogram analysis (and also before doing spectral analysis, a modified form of periodogram analysis discussed in Chapter 6).

Periodogram as Decomposition of Time-Series Variance

Consider the simulated time series example used in the previous chapters: observations of mood on 14 days. A time series of length N can be exactly reproduced by summing $N/2$ sinusoidal waveforms with cycle lengths of $N/1$, $N/2$, $N/3$, and so on, down to a cycle length of $N/(N/2)$, or two observations. The goal of periodogram analysis is to partition the Sum of Squares (SS_{total}) for an overall time series of length N into a set of $N/2$ SS components, each with 2 df, that correspond to the amount of variance accounted for by each of these cycles. (See Box & Jenkins, 1970, for a more precise description of the disposition of the degrees of freedom; the number of df allowed for the highest frequency depends upon whether N is an odd or even number. In some situations, the SS for the highest frequency has only 1 df associated with it.)

In our hypothetical example where the length of the time series is 14 days, the cycle lengths that will be fitted to the data include periods of

$$N/1, \quad N/2, \quad N/3, \quad N/4, \quad N/5, \quad N/6, \quad N/7$$
$$14/1, \quad 14/2, \quad 14/3, \quad 14/4, \quad 14/5, \quad 14/6, \quad 14/7$$

That is, the overall SS for the time series will be partitioned into the sums of squares that are accounted for by seven different cyclic components, with periods of 14, 7, 4.67, 3.5, 2.8, 2.33, and 2 days. Frequency is the inverse of period, so the set of frequencies included in the analysis consists of $1/N$, $2/N$, . . . , $(N/2)/N$, or $1/14$, $1/7$, . . . , $1/2$.

Frequency is just the proportion of a cycle that occurs during one observation. Because these frequencies are equally spaced, they are orthogonal to each other; that is, ideally, they are statistically independent of each other. The overall variance of the original time series can be divided into $N/2$ (in this example, $N/2 = 7$) SS terms (one corresponding to the goodness of fit of each of the seven frequencies).

Essentially, the same harmonic analysis that was described for the 7-day cycle in Chapter 4 is performed separately for each of these seven cyclic components, and the results are summarized in an ANOVA source table. A detailed worked example and a more mathematically oriented explanation of periodogram analysis are given by Box and Jenkins (1970, pp. 36–39). Formulas for these computations are also presented in the Appendix 5.1 at the end of this chapter. For each of the seven cyclic components, A and B coefficients are calculated to fit a sinusoid of that frequency to the time-series data. From these A and B coefficients for each frequency, we can compute the "periodogram ordinate" or intensity by taking $2/N \cdot (A^2 + B^2)$ for each frequency. This periodogram ordinate is the SS accounted for by each periodic component. Just as in a familiar ANOVA, the sum of these SS terms across all seven frequencies is equal to SS_{total} for the overall time series. Note that some computer programs (such as SAS) report the Mean Square (MS) rather than the SS for each periodic component; SPSS reports the SS for each periodogram component, which is consistent with the example given by Box and Jenkins (1970).

Note that while it is possible to obtain the estimates of the A and B coefficients for each frequency using the ordinary least squares regression computations described in Chapter 4, when the N in the time series becomes large, it is more efficient to use discrete Fast Fourier Transform (FFT) algorithms to obtain these coefficients. Most computer programs use FFT algorithms; the results are equivalent to those from the regression procedures described earlier, although the actual computational process is different (see Bloomfield, 1976, Chap. 4, for details on one popular FFT algorithm).

The null hypothesis for a periodogram analysis is that the time series is "white noise," that is, it consists of approximately equal amounts of variance due to each of the $N/2$ periodic components. If the null hypothesis were true, we would expect to see approximately equal (given sampling error) SS values across the entire periodogram. Alternatively, if one periodic component explains a large amount of variance in the time series, it will have a larger SS than would be expected by chance if SS_{total} were equally divided among the $N/2$ frequency components. A graph of the periodogram is often presented. Usually, this graph shows the intensity or SS for each frequency (plotted on either on a log or linear scale) on the Y axis and the frequencies (or sometimes the periods) on the X axis. Large peaks that correspond to periodic components that explain large proportions of the overall variance of the time series can be identified by visual examination of this graph. However, large peaks can arise by chance, so it is desirable to use statistical significance tests to assess the results.

A Significance Test for Periodogram Peaks

SPSS does not provide any tests for statistical significance of periodic components, but a significance test can easily be done using some of the results of the SPSS periodogram analysis. Some variability in the sizes of the sums of squares would be expected just due to sampling error, and if one chooses the largest SS value out of a large number of SS values in a table, it may well appear "large" even though the data are white noise. A significance test needs to take into account the inflated risk of Type I error that arises when many periodic components are examined. Many studies that use periodogram analysis or spectral analysis use qualitative judgments: when the periodogram or spectrum is graphed, "large" peaks are noted and these are described as major periodic components. A less subjective criterion for the identification of "major" periodic components is recommended here, based upon a test developed by Fisher (1929) and critical values tabled by Russell (1985). This test provides a reasonable method for testing the statistical significance of peaks in a periodogram.

To perform the Fisher test, the data analyst must first compute the test statistic g, which is the proportion of the total variance that is accounted for by the largest periodogram component. To find g, simply divide the largest SS (or intensity estimate) by the sum of the sums of squares (i.e., SS_{total} for the overall time series). The critical value for this statistic depends only on N, the length of the time series, and upon α, the risk of Type I error. A time series of length N is partitioned into $q = N/2$ periodic components; Fisher's test essentially assesses how large the largest of q such components has to be before it is unlikely that such a large peak could arise by chance in q components from white noise data. It is a "protected" or conservative test, in the sense that it takes into account the fact that the analyst is selecting the largest periodogram ordinate post hoc, rather than testing for one particular frequency specified a priori. If the largest peak is statistically significant (if g exceeds the tabled critical value), then it is also possible to go on and test the second and third largest peaks for significance, and so on. Tables of critical values of g for the first, second, and third largest peaks of time series of various lengths (from Russell, 1985) are included at the end of this book in Appendix B. For example, if the number of observations in the time series (N) is 130, then the largest peak in the periodogram must account for more than 10.722% of the variance to be judged significant at the .05 level; the second largest peak must account for more than 7.675% of the variance to be judged significant; and so forth.

Empirical Example: Periodogram Analysis of Simulated Mood Data

The worksheet for the simulated mood data set ("mood.sav") in Table 5.1 contains the periodogram results for the simulated mood time-series data described in Chapter 4. The periodogram was obtained by selecting the (graphs) option from the main menu in SPSS for Windows, then choosing the (spectra) option. Within the menu for the spectral analysis procedure, you select the periodogram option and tell the program which variable to base the periodogram analysis on. By default, this procedure produces only a graph of the periodogram or spectrum.

TABLE 5.1. Periodogram Analysis for the Mood Data, Using $N = 14$ Observations (No Leakage)

	mood	freq	per	pdg_1	pctpdg
1	4.99	.00000	.	.00	.00
2	3.37	.07143	14.00000	.21	.01
3	3.06	.14286	7.00000	11.22	.77
4	2.41	.21429	4.66667	.25	.02
5	2.91	.28571	3.50000	1.75	.12
6	3.69	.35714	2.80000	.42	.03
7	4.88	.42857	2.33333	.76	.05
8	5.35	.50000	2.00000	.00	.00
9	2.37	.	.	.	.
10	2.59	.	.	.	.
11	2.75	.	.	.	.
12	2.42	.	.	.	.
13	3.51	.	.	.	.
14	4.70	.	.	.	.

It is desirable to save the actual computed values of the periodogram so that more precise information can be read out than one could obtain just by visually examining the graph. Unfortunately, the command to save variables into the worksheet is not available on the menu; it can only be obtained by editing an additional "save" command into the spectral analysis commands in an SPSS syntax window. The (paste) command is used to insert the basic spectral analysis command lines that were generated by the menu selections of the user into a syntax window; then these commands can be edited to add a line that tells SPSS to save the computed values of the periodogram as variables in the SPSS worksheet. (See Chap. 4 in the *SPSS for Windows Base System User's Guide* for information about the use of syntax windows. See Chap. 13 in the *SPSS for Windows TRENDS Manual* (SPSS, Inc. 1993) for information on the spectral analyis procedure and a complete explanation of the options it provides through menu selections and through additional commands.)

Column 1 of the worksheet in Table 5.1 contains the simulated mood time-series data. Column 2 contains the set of frequencies that were fitted to this data; these were computed by SPSS based on an N of 14 observations and saved to the worksheet as a new variable. As shown above, the set of frequencies corresponds to $1/N, 2/N, \ldots, 7/N$. Frequency is just the fraction of a cycle that occurs during one observation. Column 3 shows the periods of the components that are being fitted to the data; these are calculated from $N/1, N/2, \ldots, N/7$. In other words, the longest cycle that will be fitted to the data is 14 days; the shortest cycle is 2 days; and seven periodic components ($N/2 = 14/2$) are included in the analysis. If the null hypothesis of white noise were correct, we would expect each of these seven periodic components to account for an equal (about 1/7) share of the overall variance of the time series. Column 4 of the worksheet contains the Sums of Squares corresponding to the goodness of fit of each of these seven periodic components, also called the periodogram intensities. The variable name SPSS assigned to this was "pdgm_1," the periodogram from model number 1.

Examination of these Sums of Squares (or periodogram intensities) indicates that the only periodogram component that has a "large" SS and that therefore accounts for a relatively large proportion of the variance in the time series corresponds to a frequency of 1/7 or a cycle length of 7 days. The SS for each frequency can be converted to a percent of the variance of the original time series by dividing it by SS_{total} for the original data (or the sum of the periodogram ordinates, which is equivalent to SS_{total}). The column in Table 5.1 that is headed "pctpdg" contains the percentages that were obtained by dividing each SS by the SS_{total}. Dividing 11.21111 by 14.58 yields .77, which agrees, within rounding error, with the earlier estimate from the harmonic analysis of the mood data (reported in Chapter 4) that a 7-day cycle accounts for 76.8% of the variance.

When this percentage of variance was compared to the critical value of g for a time series with an N of 25 (the closest available N in the table) in the Fisher test tables (B.1–B.3) in Appendix B, it was judged to be statistically significant ($p = .001$). However, the next largest SS, when converted to a percentage of variance ($1.745/14.58) = .12$) was not statistically significant using the Fisher critical values.

The periodogram is often displayed as a graph. The horizontal axis may be either the frequency or period; the vertical axis is the periodogram ordinate or SS or intensity, plotted on either a log or a linear scale. A periodogram plot is shown in Figure 5.1 (using a linear scale for the periodogram intensity). In most cases, the analyst needs more precise detail than can be read from such a graph by eye; for this reason, it is more useful to look at the periodogram results in the SPSS worksheet or in the form of a table. However, graphs similar to that shown in Figure

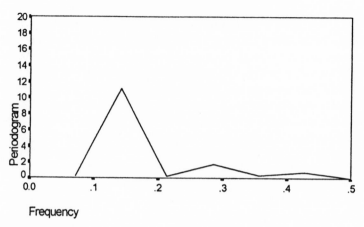

FIGURE 5.1. Periodogram of hypothetical mood data (intensity plotted on linear scale).

5.1 can be useful ways of illustrating one or several major periodic components.

Leakage: Example of a Common Problem

It is crucially important that the time-series length that is used in periodogram analysis be an integer multiple of the cycle length that we are trying to detect. (In spectral analysis, as described in Chapter 6, there is a way to modify the computation of the spectrum so that the fitted frequencies match the cycle length of interest even if the N is not an integer multiple of the cycle length of interest.) For example, if the time series consists of mood measured once a day and 7-day cycles are expected, the length of the time series should be an integer multiple of 7, such as $12 \cdot 7$ or $N = 84$. When the time-series N is not an integer multiple of the period of the cycle in the data, an artifact called "leakage" occurs (Bloomfield, 1976, p. 51). Leakage occurs when the variance actually due to some "real" cycle that cannot be accurately detected by the periodogram analysis spills over or "leaks" into the sums of squares that correspond to other cycle lengths—those that *can* be detected by the periodogram analysis.

It is easier to explain leakage by example than by formal argument. In the empirical example using the mood data, the period of interest was 7 days and the N was 14 days, so there were exactly two repetitions of the 7-day cycle. When an N of 14 was used, no leakage occurred: the periodogram accurately detected the presence of a 7-day cycle. What would happen if we had an N of only 12 days in our time series and we were trying to detect a 7-day cycle? In this case, the N (12) would no longer be an integer multiple of the cycle length of interest (7), so the set of cycles that was fitted to the data would no longer include a cycle length of 7.

A second periodogram analysis of the mood data, illustrating the occurrence of leakage, is shown in Table 5.2. In this analysis only the first 12 of the original mood time-series data points have been used. Because the N in this example is not an integer multiple of 7, note that now the frequencies included in the periodogram correspond to 1/12, 2/12, ..., 6/12 (or periods of 12, 6, 4, 3, 2.4, and 2 days). The cycle length of 7, which best fits the actual time series, is *not* included in this set. Because of this, the variance that should be accounted for by a cycle of 7 days leaks into other components; in particular, the variance due to a 6-day cycle is overestimated, because this component picks up

TABLE 5.2. Periodogram Analysis for the Mood Data, Using N = 12 Observations (Leakage Occurs)

	day	mood	period12	freq12	pdg12
1	0	4.99	.	.00000	.00000
2	1	3.37	12.00	.08333	.99867
3	2	3.06	6.00	.16667	8.27932
4	3	2.41	4.00	.25000	.07622
5	4	2.91	3.00	.33333	1.61622
6	5	3.69	2.40	.41667	1.96486
7	6	4.88	2.99	.50000	.21282
8	7	5.35	.	.	.
9	8	2.37	.	.	.
10	9	2.59	.	.	.
11	10	2.75	.	.	.
12	11	2.42	.	.	.
13	12	3.51	.	.	.
14	13	4.70	.	.	.

the variance that is really due to a 7-day cycle. The variables "freq12" and "period12" correspond to the set of frequencies and periods that we would obtain using $N = 12$. The variable "pdg12" contains the periodogram ordinates or SS terms obtained when this set of 6 frequencies is fitted to the first 12 observations of the mood time series. The largest SS component in this second periodogram analysis corresponds to a cycle of 6 days.

Were we to take the results of this second periodogram analysis too literally, we might be misled into reporting 6-day cycles in mood. This example illustrates the artifact of "leakage" (Bloomfield, 1976, pp. 51–52). If the time-series data contain a frequency that does not match one of the frequencies included in the periodogram, then the variance due to that omitted frequency will "leak" into neighboring frequencies. The variance due to a 6-day cycle is artifactually inflated (and due to a sort of ripple effect, sums of squares for 3- and 2.4-day cycles are also inflated). The point of this example is that we should not interpret the periodogram too literally. A large SS for the periodic component of 6 days suggests that there could be a cycle of approximately that length in the data, but a closer examination of the data makes it clear that this large SS for the 6-day cycle is really due to the presence of a 7-day cycle. If the cycle lengths of interest are known a priori, the data analyst can avoid this problem by making the time-series length an integer multiple of the cycle length (e.g., if we want to study 7-day cycles, our overall N should be an integer multiple of 7). If there is no prior knowledge of the cycle length, the data analyst should experiment by changing the N of the time series that is subjected to periodogram analysis to see which set of frequencies yields the clearest peaks.

The analyst should also understand that the estimation of cycle length from the periodogram is only approximate and that a large SS only suggests that there are cycles *approximately* 6 or 7 days in length. Periodogram analysis does not usually provide a precise estimate of the "true" cycle length. This need not be a problem if the research question deals with the amount of variance contained in fairly wide frequency bands, as in many of the empirical examples described in later chapters. If it is important to estimate the length of the period more precisely, then it may be useful to do harmonic analysis using a grid search to assess a range of possible values for τ to see which provides the best fit to the time series, as described in Chapter 4. If the model that best describes the time series includes more than one cyclic component, then it may be necessary to do a grid search estimation for the coefficients for each cycle using the method of "cyclic descent," as reviewed in Chapter 4, and as described in more detail by Bloomfield (1976, pp. 22–23).

Chapter Summary and a Preview: The Power Spectrum as a Modification of the Periodogram

The major criticism of periodograms is that their ordinates are subject to a great deal of sampling error or variability. Because the number of periodogram ordinates $(N/2)$ increases along with the N of the time series, merely using a larger N does not correct this problem; in all cases, the estimate of each SS is based on only 2 df. Spectral analysis refers to a family of statistical techniques used to improve the reliability of the periodogram. Essentially, a power spectrum is a periodogram that has been "smoothed" in one of many possible ways, described in Chapter 6, to make the estimates more reliable.

The advantage of using the power spectrum rather than the periodogram is that the spectrum may give a better and more reliable picture of the distribution of power (or variance accounted for) over the set of frequencies. The disadvantages of using the spectrum is that, unlike the periodogram, the spectrum does not correspond to an exact partition of the variance of the time series. Furthermore, if too much smoothing is applied to the spectral estimates, it becomes difficult to distinguish contributions of neighboring frequency bands.

Thus, for some research questions, where the researcher's goal is to assess variance partitioning or to try to get a reasonably precise first guess about cycle length to use for follow-up harmonic analysis, the periodogram may be a simpler and more useful tool than the spectrum. However, the limitations of the periodogram should be understood: the estimated "intensity" or SS for each periodic component has a great deal of sampling error associated with it; the cycle lengths that are fitted to the data by the periodogram analysis may not exactly match the cycles in the data, which can create leakage artifacts; and follow-up analyses such as harmonic anlysis may be required in order to get a more precise estimate of cycle lengths for any periodic components that are detected by periodogram analysis.

Appendix 5.1. Formulas for Computation of the Periodogram

The formal model for the periodogram is an extension of the harmonic analysis model. Instead of representing the time series as the sum of one periodic component plus noise, the periodogram model represents the time series as a sum of $N/2$ periodic components.

The model to represent a time series of N observations can be written as follows:

$$X_t = \mu + \Sigma(A_i c_{it} + B_i s_{it}), \qquad \text{for } i = 1, 2, 3, \ldots, N/2$$

where X_t is the values of the X time series; t is the time or observation number $(0, 1, 2, \ldots, N)$; c_{it} is the cosine function of frequency ω_i evaluated (at all times $t = 0, 1, 2, \ldots$); and s_{it} is the sine function of frequency ω_i evaluated (at all times $t = 0, 1, 2, \ldots$). The set of frequencies ω_i is given (in radians) by $2\pi i/N$, for $i = 1$, $2, 3, \ldots, N/2$, where $N =$ the number of observations in the time series if N is even.

SPSS assumes that N is an even number; if it is not, the length of the time series is modified by adding/dropping one observation to make it an even number. (The disposition of degrees of freedom is slightly different if N is odd; this is explained by Box & Jenkins, 1970, pp. 36–37.) Here 1 df is associated with the estimate of μ. Each of the $N/2$ periodic components have 2 df associated with them except for the highest frequency, which corresponds to a cycle just two observations long. This highest frequency periodic component has just 1 df associated with it, because it can be represented by either a sine or cosine term; separate estimates of the A and B coefficients for the cosine and sine components, in order to identify the phase, are not needed for this highest frequency.

$$\tilde{\mu} = \overline{X} = \frac{1}{n} \sum_{t=1}^{n} X_t$$

$$\tilde{A}_i = \frac{2}{n} \sum_{t=1}^{n} (X_t - \overline{X}) \cos \omega_i t, \qquad \text{for } i = 1, 2, \ldots, q$$

$$\tilde{B}_i = \frac{2}{n} \sum_{t=1}^{n} (X_t - \overline{X}) \sin \omega_i t, \qquad \text{for } i = 1, 2, \ldots, q$$

$$q = N/2 \qquad \text{(if } N \text{ is even)}$$

These equations describe the OLS estimation for the parameters of this equation (A and B); a separate estimate of a pair of A and B estimates is done for each of the Fourier frequencies. The Fourier frequencies for a given analysis are determined by the number of observations in the time series (N) and the sampling frequency (Δt). The set of periods or cycle lengths is given (in numbers of observations) by $N/1, N/2, N/3, \ldots, N/q$, where $q = N/2$. The set of frequencies (in cycles per observation) is given by $1/N, 2/N, 3/N, \ldots, q/N$. (The last term, q/N, corresponds to a frequency of .5, or a period of two observations.)

These periods can also be expressed in time units rather than in numbers of observations; to make this conversion, just substitute $N \, \Delta t$ for N in the preceding lists. For instance, if $N = 240$ observations, and $\Delta t =$ number of seconds per observation $= 10$ seconds, then the longest cycle is $N \, \Delta t/1 = 240 \cdot 10/1 = 2,400$ seconds; the shortest cycle is $2\Delta t = 2 \cdot 10 = 20$ seconds long.

The Fourier frequencies are chosen such that the longest cycle that is fitted to the time series is equal to the length of the time series. The shortest cycle has a period of two observations. The frequencies in between are equally spaced, and they are orthogonal to each other, that is, they account for nonoverlapping parts of the variance of the time series.

For each of these q frequencies (except the highest frequency, which only needs one parameter), we can compute an A and a B coefficient using the same equation that was used earlier to estimate coefficients in harmonic analysis in Chapter 4. Essentially, we multiply the deviations from the mean of the observed time series by the cosine and sine functions of a particular frequency, and obtain estimates of the A and B coefficients that provide the best fit for that waveform to the data. We can summarize goodness of fit by computing an "intensity" or periodogram ordinate or SS for each frequency: the periodogram ordinate for frequency i is $(N/2) \cdot (A_i^2 + B_i^2)$. These periodogram ordinates can be reported in the form of a table, or they can be graphed as a function of frequency or period. The analyst is interested in seeing which, if any, of the periodic components account for large proportions of the variance in the time series.

Periodogram analysis can be understood essentially as a one-way ANOVA in which the SS_{total} for the overall time series is partitioned into q Sums of Squares, each of which assesses the goodness of fit of one of the q periodic components to the time-series data. The computations described so far are essentially OLS regression analyses where the predictor variables happen to be variables that represent cosine and sine waveforms of different frequencies; and in fact OLS computational methods can be used to obtain the A and B coefficients to fit a cosine and sine waveform of any specific period to an observed time series (Bloomfield, 1976, pp. 43–44). However, if the time series is long (and correspondingly, if the number of periodic components that are fitted to the data is large), then this straightforward OLS estimation approach becomes computationally inefficient.

Alternative computational methods to obtain the A and B coefficients for the set of Fourier frequencies are therefore used by most computer programs. Most of these are based on the discrete Fourier transform and computationally efficient algorithms that are generally referred to as Fast Fourier Transforms (FFT). While it is not essential for a data analyst to understand this approach, a brief intuitive explanation is provided here. The mathematical details are provided in Bloomfield (1976, Chap. 4).

The Fourier transform makes use of a trigonometric identity called the Euler relation:

$$e^{i\lambda} = \cos \lambda + i \sin \lambda$$

This equation essentially says that we can decompose a time series variable λ into cosine and sine components by taking $i\lambda$ and raising e to that power. In this equation, e is the constant that is the basis for natural logarithms, and i is the imaginary quantity the square root of -1. The mechanics for computation of the discrete Fourier transform are complex, and the equations for the Fourier transform are not included here (see Bloomfield, 1976, Chap. 4, for technical details). Essentially, when a discrete Fourier transform is applied to a set of time-series data, the result is a set of complex numbers. A complex number is of the form $a + bi$, where a is the "real" part and b is the "imaginary" part of the complex number. It happens that the a and b components of the complex numbers obtained from the Fourier transform correspond exactly to the A and B coefficients for the sine and cosine functions in a harmonic analysis; the Euler relation shows that this discrete Fourier transform is just another clever way of splitting a time series into its sine and cosine components. The discrete Fourier transform for a time series with N observations consists of $N/2$ complex numbers; each of these numbers has a "real" part that is equivalent to the coefficient for the cosine component of one of the Fourier frequencies (which was earlier denoted as A), and an "imaginary" part that is equivalent to the coefficient for the sine component of one of the Fourier frequencies (earlier denoted as B).

Here A_i and B_i are the coefficients that fit a sinusoidal waveform of frequency i to the time series data, and from these A and B coefficients we can compute an amplitude, a phase, and an SS accounted for by each periodic component.

There is an extensive literature on efficient algorithms for performing the FFT. One example is the Cooley–Tukey algorithm (Bloomfield, 1976, Chap. 4). Some algorithms require particular time-series lengths (e.g., N must be an even number for some algorithms; N must be a power of 2 for some algorithms) in order to employ computational shortcuts that greatly reduce the amount of computation that is required. It is not necessary for data analysts to know much about these computational algorithms except to understand that the particular algorithm used by the computer program may require a particular time-series length.

Some writers use the terms "FFT" and "periodogram" loosely and almost interchangeably, which can create confusion. In this book the term "periodogram" always refers to the set of sums of squares or intensity estimates that are obtained by taking $(N/2) \cdot (A_i^2 + B_i^2)$ for each of the frequencies. These can be summarized either in a graph (of periodogram intensity against frequency) or in a table. In this book, FFT always refers to the set of A and B coefficients for each frequency that correspond to the complex numbers obtained from the Fourier transform (or equivalently, these same coefficients can be obtained from the OLS estimation procedures described in Chapter 4 when harmonic analysis was introduced).

The reader should note that some computer programs and textbooks define the periodogram slightly differently. For instance, whereas SPSS reports the SS for each frequency component, which is consistent with the definition of the periodogram used here and in Box and Jenkins (1970), SAS reports the MS for each frequency when the periodogram is requested. If it is not clear whether the program is reporting Sums of Squares or Mean Squres, an easy way to check this is to input a small data set that is used as an example in a textbook—such as the mood time series used as an example in this book—to see whether the output from the program agrees with the results in the text. If a progam produces the same numbers shown in the column headed "pdg_1" in Table 5.1, then the program has reported Sums of Squares; if it reports values that are one half the size of the numbers in this table, Mean Squares have been reported.

A great advantage of SPSS for Windows is that it permits the analyst to retain many of the intermediate results in the computation of the periodogram or spectrum as new variables. Further computations can be performed using these variables. For instance, the estimates of the periodogram Sums of Squares for each Fourier frequency can be retained as a variable; this makes it simple to find an estimated proportion of variance accounted for by each periodic component. In addition, SPSS makes it possible to save the A_i and B_i coefficients for each frequency (the FFT complex number coefficients—or, equivalently, the coefficients for the harmonic analysis results for each individual periodic component). As demonstrated in Chapter 4, these coefficients can be used to reconstruct the sinusoidal components that correspond to peaks in the periodogram, and then to plot these reconstructed periodic components against the original time series. These plots provide a helpful illustration showing which features of the time series have been captured by the analysis.

CHAPTER 6

Spectral Analysis

Introduction

Periodogram analysis, as described in Chapter 5, can be a useful way of assessing whether there is a strong cyclic component in a time series. However, its most serious fault is that the sampling errors associated with estimates of Sums of Squares are quite large; confidence intervals set up around the "intensity" estimates at each frequency are therefore quite wide. Spectral analysis techniques were developed to reduce this problem of sampling error. A power spectrum is a slightly modified version of a periodogram; there are many different versions of spectral analysis that involve different ways of modifying the periodogram estimates to reduce their sampling error.

This chapter describes several common methods for deriving a power spectrum from a periodogram. The term "smoothing" refers to a process in which each periodogram intensity is replaced by a weighted average that includes intensity estimates for a few neighboring frequencies. Smoothing procedures differ in two ways: First, the width of the "window," that is, the number of neighboring frequencies that are included in this weighted average, can vary. Second, the weights used for this weighted average can be of different forms; some smoothing windows give equal weight to all included frequencies, whereas others give more weight to frequencies near the center of the window than to frequencies near the edges. When the weighting function is graphed, it may have various different shapes; for instance, a Daniell window looks like a rectangle, whereas many of the other popular smoothing windows have a bell shape. Thus, windows can vary in width and also in shape.

Before smoothing, the graph of a periodogram is typically quite jagged and the confidence intervals around each intensity or sum of squares estimate are quite wide. After smoothing, the resulting power spectrum is typically somewhat flatter; neighboring peaks that were separate in the periodogram sometimes become combined into one relatively broad peak; and the confidence intervals around the estimates are narrower, because the smoothing tends to reduce sampling error.

This chapter describes some of the more popular methods of smoothing. It describes how to estimate the "equivalent degrees of freedom" for one of these windows (Daniell), and how to use the χ^2 distribution with these equivalent degrees of freedom to set up confidence intervals around each spectral estimate. Guidelines for identification of major periodic components from a power spectrum are suggested.

A power spectrum is a periodogram that has been smoothed, using one of many possible smoothing functions, or "windows," to reduce the sampling error. A periodogram partitions the variance of the overall time series into a discrete set of frequency components; the sum of squares associated with each frequency was called an "intensity." In a spectrum these sums of squares are averaged together across neighboring frequencies to provide a smoother and more reliable estimate of the distribution of variance continuously across the entire range of frequencies from $1/N$ to ½. The term "power" is typically used to refer to the estimated amount of variance in the time series that is accounted for by a particular band of frequencies.

As with the periodogram, it may be useful to standardize the power spectrum by dividing each spectral estimate by the overall amount of power or variance to yield spectral density. Spectral density provides an estimate of the proportion of the variance in the time series that is accounted for by a particular frequency band. For some purposes, spectral density may be more comparable across subjects than the raw power spectrum, where the sizes of peaks are a function of both the distribution of power across frequencies and the overall amount of variability in the subject's time-series data.

Description of Commonly Used Windows for Smoothing Spectra

There is a vast literature on smoothing procedures, or "windows." Only a brief and nontechnical introduction is provided here; the reader is advised to consult more technical sources, such as Bloomfield (1976), Gottman (1981b), or Koopmans (1995) for more details and more technical treatment of the subject.

The SPSS SPECTRA program offers a menu of different windows, including Hamming, Bartlett, Parzen, Tukey, and Daniell (also called the equal-weight or "unit" window in the SPSS SPECTRA program). For specifics on the weighting functions that are used for each type of smoothing, see pp. 300–301 in the *SPSS TRENDS Manual*. In this chapter, the Daniell or equal-weight window will be used as the smoothing procedure because of its simplicity.

A smoothing window is defined as a set of weights that are used to compute weighted averages; each periodogram intensity estimate is replaced by a weighted average of intensities within a few neighboring frequencies. As defined by most textbooks (Gottman, 1981b; Jenkins & Watts, 1968; Koopmans, 1995), these weights are usually normalized to sum to 1, for the sake of convenience. However, the size of the weights is sometimes implemented differently by computer programs, and therefore it is important to examine the actual weights used by the program instead of relying on general descriptions of a type of window from reference books, which may or may not agree with the specific procedures used by the program.

A window has two characteristics. The first is its *width:* the total number of neighboring frequencies that are included in the weighted average. Because the window is symmetrical around some center frequency, the total width of the window (denoted M) is usually an odd number; usually the intensity estimate for one frequency is averaged together with equal numbers of frequencies (m) above and below it. Thus $M = 2m + 1$, where m is the number of terms on each symmetrical half of the window. The second characteristic is its *shape:* the nature of the weighting function that is used to compute a weighted average of intensities at neighboring frequencies. For example, the simplest window—the Daniell or equal-weight window—gives equal weights to the periodogram ordinates that are averaged together. However, other windows use different weighting functions; most windows, such as the Hamming or Hanning windows, give larger weights to the frequencies near the center of the window than to frequencies near the edge of the window. When graphed, the Daniell window corresponds to a rectangular or "boxcar" function. When graphed, many other windows are tapered toward the edges and resemble a cosine waveform or a bell shape.

In the following descriptions of specific windows, each window involves computing a weighted average of intensities of sums of squares at "neighboring" frequencies. Recall that the set of Fourier frequencies is given as follows: $1/N, 2/N, 3/N, \ldots, 1/2$. In general, the frequency f_i for periodic component i is i/N, for $i = 1, 2, \ldots, N/2$. Thus, the frequencies that are "neighbors" to the frequency f_i are $f_{i+1}, f_{i+2}, \ldots$ and $f_{i-1}, f_{i-2}, \ldots$.

A Daniell window of width $M = 3$ involves replacing the intensity estimate for frequency i, with the mean of the intensity estimates for frequencies $i - 1$, i, and $i + 1$. According to most textbooks, using the Daniell window, the smoothed estimate of the power at frequency i in the power spectrum is obtained by taking $(1/3) \cdot$ periodogram intensity at frequency $(i - 1) + (1/3) \cdot$ intensity at frequency $(i) + (1/3) \cdot$ intensity at frequency (i + 1). (Actually, SPSS SPECTRA simply sums these three values instead of averaging them.) That is, the three neighboring periodogram ordinates are combined together (equally weighted) to create a more reliable estimate of a continuous spectrum. The width of the window (3) is the number of frequencies included in the weighted average.

Note that virtually every source book describes a Daniell window of width n as having a weight of 1/n for each term included in the weighted average. However, the SPSS SPECTRA program uses unit weights (instead of weights of 1/M) when a Daniell window of width M is requested. That is, in SPSS, a Daniell window of width M = 5 replaces each periodogram estimate for frequency i with the sum of five periodogram estimates (for frequencies i − 2, i − 1, i, i + 1 and i + 2). Thus the power spectrum created by SPSS SPECTRA using a Daniell window tends to have much larger values than the periodogram because it is based on a *sum* of components rather than a weighted average.

Generally, the wider the window, the more the smoothing process will tend to flatten the periodogram. Also, the flatter the shape of the window, the more the smoothing process will tend to flatten the periodogram. Depending upon the window that is used for smoothing a periodogram, the resulting power spectrum may be much "flatter" than the original periodogram. Two neighboring peaks in the periodogram may merge into one broad peak in the spectrum after smoothing.

Formulas that describe the weighting functions for a variety of other widely used smoothing windows are presented in detail in many other sources, including Gottman (1981b) and Koopmans (1995). Technical issues concerning the relative advantages and disadvantages of using these windows are presented in much detail by these authors. The recommendation made here is that the data analyst try out several windows in the early exploratory stages of data analysis to see how various types of smoothing affect the shape of the estimated spectra.

The trade-off involved in smoothing is the following: The more the periodogram ordinates are smoothed, the more the sampling error that is associated with the resulting smoothed spectral estimates is reduced. In fact, the effective or equivalent degrees of freedom (*edf*) associated with the estimate of power at each frequency is increased by smoothing (see Koopmans, 1995, Chap. 8, for details). If one wishes to set up confidence intervals around spectral estimates (which can be done using procedures outlined by Koopmans, 1995, on pp. 275–277), the wider the smoothing window, the narrower the confidence interval will be for the resulting smoothed estimate.

However, the more the periodogram is smoothed, the more difficult it becomes to distinguish between the contributions of neighboring frequencies or to make judgments about the period or cycle length of major periodic components. In some cases, an initial periodogram shows two or three separate peaks; after smoothing, these peaks may be blurred into one wider and flatter peak in the spectrum. This makes it difficult to answer the question whether there really are three distinct and separate periodic components in the time series.

An Alternative Approach to Smoothing:
The Bartlett Window

The smoothing procedures, or windows described in the previous section involve taking the Fast Fourier Transform (FFT) or periodogram of a time series and then averaging together neighboring periodogram intensities using a weighting function called a window to generate a smoother, more reliable estimate of the distribution of power across frequencies called a power spectrum. An alternative approach to estimation of a spectrum is to calculate the lagged autocorrelation function (ACF) for a time series up to some lag M; then do an FFT or periodogram analysis on this lagged ACF. This method is called the Bartlett window. The set of Fourier frequencies that is fitted to the data is now based on M, the maximum lag in the ACF, instead of N, the number of observations in the original time series. A benefit of this approach is that the analyst can choose various values of M as a means of fitting different sets of periods to the data. This can be a way of avoiding the problem of leakage when N is not an integer multiple of the cycle length that the analyst is trying to detect, making it possible to vary the set of periods or Fourier frequencies that are fitted to the time series, while still making use of all the data.

Note that this method of obtaining the spectrum (first computing the ACF to lag M, then applying the FFT) is the procedure that Koopmans (1995, p. 275) calls the Bartlett (1) window. For the resulting spectral estimates in this case the $edf = N/M$, where N is the number of observations in the original time series and M is the maximum lag in the ACF that is used as the basis for the FFT. The "Bartlett" window that is provided by the SPSS SPECTRA program is a different procedure.

Recommendations on Window Selection

In practice, the choice of the best window for smoothing is a matter of subjective judgment. A data analyst may try several different window widths or shapes to see how different amounts of smoothing alter the shape of the spectrum. If too much detail about the shape of the spectrum is lost, the analyst may decide that it has been smoothed too much and may elect to use a narrower window in order to preserve more detail, or may prefer to use the umsmoothed periodogram as a description of the cycles in the time series. Readers who are interested in more technical detail about the way window shape and window width affect reliability and the ability to distinguish neighboring frequency bands can find this in Bloomfield (1976), Gottman (1981b, Chap. 17), Koopmans (1995), and other sources.

The Daniell window has the virtue of simplicity; however, it can produce distortions in the smoothed spectrum, such as "ghost peaks" that do not correspond to any real periodicity in the time series (Gottman, 1981b, p. 219). A better choice for most smoothing may be the Tukey–Hanning window, which does not tend to produce such artifacts. However, it is easy to determine the *edf* for a Daniell window and more difficult to determine the *edf* for the Tukey–Hanning and other bell-shaped smoothing windows. For this reason, if the data analyst wants to be able to set up confidence intervals around the power spectrum, the Daniell window is a more convenient choice; it will be the one used for smoothing in the empirical example later in this chapter.

If a researcher's question is about variance partitioning, then a good argument can be made that the periodogram may be a more useful result to report than a smoothed spectrum. In my own research, I have often used the percentage of variance contained within certain frequency bands in the periodogram as an index of "cyclicity" (Warner, Malloy, Schneider, Knoth, & Wilder, 1987; Warner, 1992b). The relatively poor reliability of the individual periodogram estimates can also be improved by averaging standardized periodograms together across many subjects (as in Warner & Stevens, 1991). However, if the analyst has only one time series and if statistical significance testing is of great importance, then smoothing may be a useful way of improving the reliability of the spectral estimates.

Significance Testing of Peaks in Power Spectra

Significance testing of spectral estimates can be performed using methods outlined in Koopmans (1995, Chap. 8). The upper and lower bounds of a confidence interval around each spectral estimate can be computed using a χ^2 distribution. In order to set up this confidence interval, it is necessary to determine the equivalent degrees of freedom (*edf*) for the new (smoothed) spectral estimates. Prior to smoothing, each periodogram intensity estimate has 2 *df*; after smoothing, the estimated new *edf* will vary depending upon the width and shape of the smoothing window. For the Daniell window, the *edf* is easily determined: it is just 2M, where M is the total width of the window (the number of frequencies that are included in the weighted average).

Computation of the weights and of the *edf* for other windows is more complex and requires numerical approximations that will not be described here; for details, see Gottman (1981b, Chap. 17 and appendix) and Koopmans (1995, Chap. 8).

For a 98% confidence interval (CI), the critical values of χ^2 (for this *edf*) that cut off the bottom and top 1% are obtained by table lookup in the χ^2 distribution. In the formula that follows, these are denoted $\chi^2_{.01}$ for the bottom 1% cutoff and $\chi^2_{.99}$ for the top 1% cutoff critical values of χ^2. The spectral estimate for which the CI is to be set up is denoted $s(f_i)$, where f_i is the specific frequency and $s(f_i)$ is the estimated power at that frequency. Then, based on formulas given by Koopmans (1995, pp. 274–276), the upper and lower bounds of a 98% CI for $s(f_i)$ are given by the following:

$$\text{Lower bound of CI} = [edf \cdot s(f_i)]/\chi^2_{.99}$$

$$\text{Upper bound of CI} = [edf \cdot s(f_i)]/\chi^2_{.01}$$

The width of the CI varies as a function of the size of $s(f_i)$, and so if $s(f_i)$ is plotted on a linear scale a separate CI must be set up for each spectral estimate. If $\log(s(f_i))$ is used, then the width of the CI is constant (see Koopmans, 1995, p. 275). This uniform width of the CI may be more convenient in some applications.

Given that the null hypothesis ("white noise") implies a uniform distribution of power across frequencies, one can draw a line on the graph of the power spectrum that corresponds to this mean level of power (i.e., the mean of all the spectral estimates). To test whether a specific spectral estimate is statistically significant, one can then examine its CI to see whether it overlaps this mean. Should it *not* overlap the mean of the spectrum, then this spectral estimate might be judged significantly higher than expected by chance. However, this approach to significance testing raises a problem of inflated Type I error (Bloomfield, 1976, p. 197). If the analyst specifies a very limited number of frequencies a priori and tests whether the power at these frequencies significantly exceeds the mean level of power (that would be expected if the series were white noise), this significance testing procedure may have acceptably low risk of Type I error. When the largest peak in a spectrum is tested in this manner post hoc, then the actual risk of Type I error will be higher than the nominal α level.

Because the CI approach to significance testing does not control for the inflated risk of Type I error that arises when many spectral estimates are tested post hoc for significance, the analyst who wishes to take a more conservative approach may instead apply the Fisher test to the periodogram as a means of assessing significance of periodicity (see Chapter 5). Alternatively, the researcher may wish to set up a CI for 99% or 99.9% to reduce the risk of Type I error.

Empirical Example: Spectral Analysis of Blood Pressure Time-Series Data

To illustrate spectral analysis in action, consider a spectral analysis of a real data set where the periods were not known a priori. Systolic blood pressure (SBP) was measured once every 2 seconds using a noninvasive finger cuff (Finapres). The subject was a male college student; he was observed during a get-acquainted conversation with a female college student. Previous research suggested that cycles on the order of 3–6 minutes may occur in SBP (Benton & Yates, 1990; Hlastala, Wranne, & Lenfant, 1973). A 40-minute session was run in order to observe up to 12 or 13 repetitions of this 3-minute cycle. Because the cycles of interest were relatively long compared to the sampling frequency of 2 seconds for the original data, for this analysis the SBP measurements were aggregated or averaged into 10-second time blocks. The time series shown at the end of this book in Appendix A in the SPSS worksheet entitled "sbp.sav" gives the mean SBP reading for this subject for each 10-second time block. (Each measure reported in this worksheet is a mean of five readings taken from the Finapres monitor.) An arbitrarily selected section of the 40-minute data record that was collected is reported here: 128 observations (or 1,280 seconds) is included.

A graph of this time series (SBP vs. observation number) is shown in Figure 6.1. Visual examination of this graph suggests possible periodicity: note the sudden drops in SBP that are seen at about 370, 500, 700, 860, and 1,000 seconds. The periodicity is much less clear cut than in the examples in earlier chapters (the 7-day cycles in the simulated mood time series and the 12-month cycles in numbers of airline passengers).

Visual examination of the graph did not suggest the presence of a linear or curvilinear trend. However, an OLS trend model was fit using regression to predict observed SBP at time t from the observation number. Because the linear trend was nonsignificant, no trend removal was applied to these data prior to doing periodogram and spectral analysis.

Both periodogram analysis and a spectral analysis were performed, using the Daniell (or equal-weight) window with a width of M = 5 to smooth the spectrum. Results were plotted, and saved as variables into

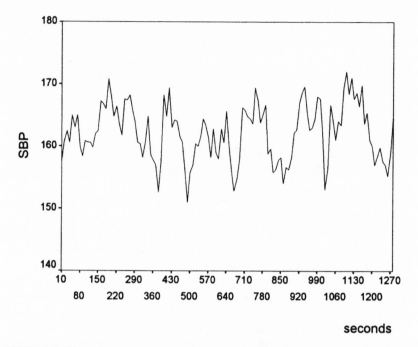

FIGURE 6.1. Plot of raw time-series data: systolic blood pressure (SBP). SBP was measured once every 10 seconds for 1,280 seconds; $N = 128$ observations.

an SPSS worksheet. See Table 6.1 for an excerpt from this SPSS worksheet. Because there were a large number of variables, this table was continued on to a second page. Only the values for the low-frequency end of the spectrum (for periods that range from 1,280 down to 64 seconds) were included, as there were no significant cycles with shorter periods in this data set.

SPSS reports frequencies in terms of number of cycles per observation. For an N of 128 observations, the Fourier frequencies consist of the following set of equally spaced frequencies: 1/128, 2/128, 3/128, ... ½ (these frequencies are expressed in number of cycles per observation). These appear in the column headed "freq" in the SPSS worksheet in Table 6.1. The corresponding cycles (with cycle length expressed as number of observations) would be 128, 64, 42.67, ..., 2 observations long. Because each observation in this data set corresponds to a 10-second time interval, the cycle lengths in seconds are 1280, 640, 426.67, ... 20. These periods (in seconds) are given in the column headed "period" in the SPSS worksheet in Table 6.1; they were calculated by multiplying the periods that SPSS provided in number of observations by 10

TABLE 6.1. Excerpt from SPSS Worksheet Showing Periodogram and Spectral Analysis Results for Systolic Blood Pressure Data (Only for Cycles Ranging from 1,280 to 64 Seconds)

seconds	sbp	freq	period	pdg_l	pctpdg	spec	a	b	c427	c183
10	157.60	.0000	.	.00000	.0000	386.89	162.21	.00	-2.19	2.89
20	160.80	.0078	1280.00	111.11951	.0432	611.31	1.23	-.48	-2.15	3.23
30	162.40	.0516	640.00	82.32339	.0320	593.61	-1.12	.19	-2.07	3.19
40	160.60	.0234	426.67	306.74425	.1194	634.10	-2.17	-.25	-1.95	2.78
50	165.00	.0313	320.00	93.42358	.0364	628.91	-1.08	.54	-1.78	2.04
60	163.00	.0391	256.00	40.48904	.0158	1227.52	-.15	-.78	-1.57	1.07
70	165.00	.0469	213.33	105.92637	.0412	1010.84	-.41	1.22	-1.33	-.03
80	159.80	.0547	182.86	680.93543	.2650	956.20	2.21	2.39	-1.06	-1.13
90	158.40	.0625	160.00	90.06150	.0351	977.54	-.18	-1.17	-.77	-2.09
100	160.80	.0703	142.22	38.78986	.0151	922.98	.77	.13	-.46	-2.81
110	160.60	.0781	128.00	61.82573	.0241	258.83	-.07	.98	-.14	-3.21
120	160.60	.0859	116.36	51.37200	.0200	205.82	-.89	.13	.18	-3.22
130	159.80	.0938	106.67	16.77778	.0065	247.12	.30	-.42	.50	-2.86
140	162.00	.1016	98.46	37.05812	.0144	189.91	-.41	.64	.80	-2.17
150	162.40	.1094	91.43	80.08810	.0312	140.25	.06	-1.12	1.09	-1.22
160	167.20	.1172	85.33	4.61654	.0018	188.60	.26	.05	1.36	-.13
170	166.80	.1250	80.00	1.71311	.0007	241.85	-.10	.13	1.60	.98
180	166.00	.1328	75.29	65.11923	.0253	166.51	-.15	1.00	1.80	1.97
190	170.80	.1406	71.11	90.31481	.0352	201.37	-.24	-1.16	1.96	2.73
200	168.00	.1484	67.37	4.74803	.0018	210.85	-.24	-.13	2.08	3.17
210	164.80	.1563	64.00	39.47344	.0154	264.27	.64	.46	2.16	3.24

Note. The variables "c427" and "c183" appear as C_{427} and C_{183} in the text equations; "sbp" denotes systolic blood pressure; the other variables are explained in the text.

(the number of seconds per observation). In general, if a single observation consists of Δt time units, then the cycle lengths can be converted from cycle length in observations to cycle length in time units by multiplying by Δt, the number of time units per observation. Because there were $N = 128$ observations in this time series, a set of 64 ($N/2$) frequencies is the basis for the periodogram and the spectrum.

If the "white noise" null hypothesis were true, we would expect each of these 64 periodic components to account for approximately 1/64, or .0156, of the overall variance of the time series. Therefore, as evidence against this white noise null hypothesis, we are interested in looking for large peaks in the periodogram and spectrum that correspond to frequencies explaining a much larger proportion of the variance in the time series.

The (unsmoothed) periodogram for this SBP time series is shown as a graph in Figure 6.2. The SPSS worksheet in Table 6.1 also contains the periodogram intensity estimate for each frequency (in the column headed "pdg_1"). Visual examination of the column of periodogram intensities (or sums of squares) reveals two relatively large peaks that correspond to cycles of about 427 and 183 seconds. It is easier to determine

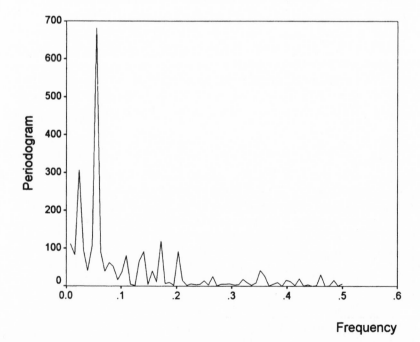

Frequency

FIGURE 6.2. Periodogram intensities for SBP (plotted on linear scale).

exactly which frequencies these correspond to by looking at the tabled values in the SPSS worksheet than by trying to read these values from the graph. Note that the two largest periodogram intensity values in the column headed "pdg_1" are 680.935 (corresponding to a period of about 183 seconds) and 306.744 (corresponding to a period of about 427 seconds). In other words, the periodogram suggests that cycles about 3 and 7 minutes long, respectively, account for a relatively large share of the variance in this time series.

Because we do not know exactly what periods we are looking for, the "real" periods in the data may not closely match any of the periods included in the set of Fourier frequencies. The occurrence of leakage is a possibility, and this may mean that the estimated cycle lengths derived from examination of this periodogram are not very accurate. All we know from Table 6.1 is that we have inflated power in the neighborhood of these two frequencies; we may be able to pin down the cycle length more precisely with more detailed follow-up analyses, but at this point we can only make approximate statements about the periods or cycle lengths.

In order to estimate what percentage of variance these two cyclic components accounted for, the periodogram was converted to percentage estimates. Each periodogram intensity estimate was divided by the sum of the all the periodogram intensities (i.e., the total of all the values in the column headed "pdg_1"). The resulting proportions are listed in Table 6.1 in the column headed "pctpdg." Examining this column, we find that the cycle about 183 seconds (or some 3 minutes) long accounted for about 26.5% of the variance in the time series; the cycle about 427 seconds (or some 7 minutes) long accounted for about 11.9% of the variance in the time series. These variance estimates can be summed to yield an overall explained variance of about 38.4% for these two cyclic components combined.

The Fisher test (described in Chapter 5) was used to test the statistical significance (α = .05) of these two peaks. For the largest peak, the obtained value of g (i.e, the percentage of variance explained by this cyclic component) was .265; this exceeds the critical value of g for n = 130 (from the table of critical values for the Fisher test, in Appendix B, this critical value was .10722). For the second largest peak, the obtained value of g was .119, which also exceeded the critical value for the second largest peak (from the table in Appendix B, the critical value was .07675). None of the remaining peaks exceeded the critical value (.06370) for a third largest peak, so only the first two peaks were judged to be statistically significant.

A spectral analysis was also done; that is, the periodogram intensity estimates were smoothed to yield an estimated power spectrum. After smoothing using a Daniell or equal-weight window of width 5, the spectrum (plotted in Figure 6.3) was somewhat smoother and flatter than the periodogram. One relatively broad peak occurred at the low-frequency end of the spectrum (the center of this peak corresponded roughly to a frequency of .039 cycles per observation, or a cycle length of about 256 seconds). The two peaks that were seen in the periodogram (corresponding to cycles of about 186 and 427 seconds) became "blurred" by smoothing and no longer stand out clearly as separate features. The numerical values for this smoothed spectrum are shown in the column of Table 6.1 that is headed "spec."

Significance testing can be done on this peak in the smoothed spectrum using the procedures outlined by Koopmans (1995), described earlier. The equivalent degrees of freedom (*edf*) for the Daniell window with a width of M = 5 is given by 2M, where M is the width of the window. Thus, *edf* = 10. To set up a 98% CI, we need to look up the χ^2 distribution with 10 *df* and find the critical values of χ^2 that cut off the bottom 1% of the area (critical value = $\chi^2_{.01}$ = 2.558) and the top 1% of the area (critical value = $\chi^2_{.99}$ = 23.209. The largest value in the spectrum is $s(f_i)$ = 1227.52 (this estimates the amount of power associated with a cycle about 256 seconds long). Substituting these values into the equations given earlier, we obtain the following lower and upper bounds for the CI around this value in the spectrum:

Lower bound = $edf \cdot s(f_i)/\chi^2_{.99}$ = (10 · 1227.52)/23.209 = 528.89

Upper bound = $edf \cdot s(f_i)/\chi^2_{.01}$ = (10 · 1227.52)/2.558 = 4798.75

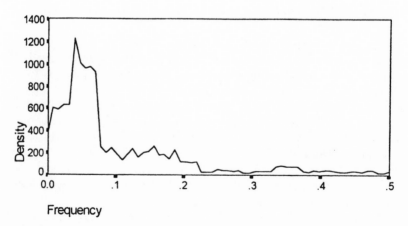

FIGURE 6.3. Spectral density of SBP using Daniell smoothing; width of window = 5; spectral density plotted on a linear scale.

The mean for the spectrum (i.e., the mean for all the values in the column of Table 6.1 headed "spec" that contains all the spectral estimates) was 448.68. Because the lower bound for the CI around this largest spectral value is higher than this mean, we can conclude that this peak is statistically significant. However, this is not, like the Fisher test, a "simultaneous" confidence interval that is protected against inflated Type I error; one would expect, in a set of 64 spectral estimates to have $64 \cdot .02$ (about 1) values that would be significant using a 98% CI as criterion. The Fisher test for the largest periodogram ordinate is more conservative.

Based upon results like these, the data analyst needs to make a provisional decision: should we believe the periodogram result (which suggests two cyclic components with periods of 183 and 427 seconds) or the spectral analysis results (which suggest one cyclic component with a cycle length around 256 seconds)? Actually, it would be a mistake to believe that either result is a precise description of the behavior of the time series. Both results are merely suggestive of a pattern. It is important to go back and look at the original time series to assess whether there really do seem to be regularly recurring cycles with these periods. Past research can also be helpful in deciding which analysis provides a better description. Many studies of cycles in blood pressure have found cycles that are about 3 and 6 minutes long (Benton & Yates, 1990; Hlastala, Wranne, & Lenfant, 1973; Kushner & Falkner, 1981; Warner, Waggener, & Kronauer, 1983). This corresponds fairly closely to the periods of 183 and 427 seconds found in the periodogram for the SBP data. On the basis of the consistency with past research, the periodogram results (which suggested two separate cyclic components) were used as the basis for further data exploration and analysis instead of the spectral analysis results, which suggested one cyclic component.

To reconstruct the two cyclic components that were identified by the periodogram analysis and assess how well they fit the original time series data, two sinusoidal functions were generated using the periodogram results and plotted against the original time-series data. The FFT coefficients, that is, the A and B coefficients (the complex number coefficients that are applied to the cosine and sine functions in the discrete Fourier transform) for each frequency component were saved as variables named "a" and "b" in the SPSS worksheet in Table 6.1. These coefficients can be used to reconstruct the two cyclic components that were identified as major periodic components from the periodogram analysis. Based on the periodogram, two periodic components were chosen: the ones that corresponded to periods of 427 and 183 seconds. The A and B coefficients for these two periodic components were found in the column of the SPSS worksheet that contains all these coefficients. SPSS compute statements were then used to create new variables that reconstruct these two periodic components that correspond to cycles about 427 and 183 seconds long.

For the 427-second cycle, here is the equation that represents the fitted sinusoidal function, using the "a" and "b" coefficients from the FFT or periodogram analysis (see Table 6.1):

$$C_{427} = A_{427} \cos(2\pi t/427) + B_{427} \sin(2\pi t/427)$$

$$C_{427} = -2.175 \cos(2\pi t/427) - .251 \sin(2\pi t/427)$$

For the 183-second cycle, here is the equation that represents the fitted sinusoidal function:

$$C_{183} = A_{183} \cos(2\pi t/183) + B_{183} \sin(2\pi t/183)$$

$$C_{183} = 2.215 \cos(2\pi t/183) + 2.395 \sin(2\pi t/183)$$

The values computed for these new variables are reported in the SPSS worksheet in Table 6.1. The mean of the original time series was added to each of these cyclic components to adjust the level so that it would be the same as the original time series (see Figure 6.1); these new variables are shown plotted superimposed on the original time series in Figures 6.4 and 6.5. A composite fitted function that consists of the

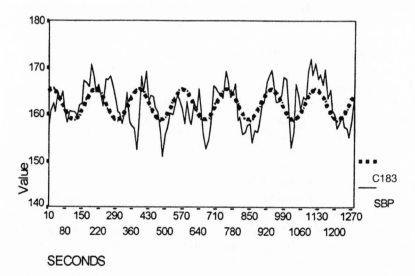

FIGURE 6.4. The 183-second cycle, reconstructed from FFT coefficients, superimposed on SBP time-series data.

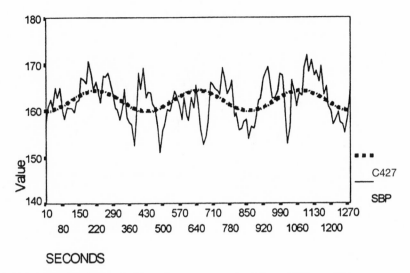

SECONDS

FIGURE 6.5. The 427-second cycle, reconstructed from FFT coefficients, superimposed on SBP time-series data.

mean of the time series, plus the 183-second cycle, plus the 427-second cycle is shown superimposed on the raw time-series data in Figure 6.6.

Visual examination of these graphs suggests that the original time series does show (somewhat irregular) cycles that are reasonably well described by sinusoids with periods of 183 and 427 seconds. Figure 6.6 suggests that what is being detected by the periodogram are the fairly regular, and rather abrupt, drops in SBP that were noted earlier based on preliminary visual inspection of the graph. These drops are spaced about 20 observations (about 200 seconds) apart. These do not occur in the first 30 observations, but they seem to occur fairly (but not perfectly) regularly after that. The relatively sudden drops in SBP do seem to occur at approximately 183-second (or 3-minute) intervals throughout the entire data record, except for the first few minutes. Thus, the impression given by the periodogram (that there are cycles about 3 and 7 minutes long) seems to be confirmed by a closer look at the original time-series data.

The percentage of variance explained by these components (derived from the periodogram) is also equal to r^2, the squared Pearson correlation between these reconstructed cyclic components and the original time series. In other words, the percentage of variance due to a 183-second cycle is just the r^2 between the 183-second cycle with the best-fitting amplitude and phase and the original time series. This can be confirmed easily by simply running Pearson correlations between the recon-

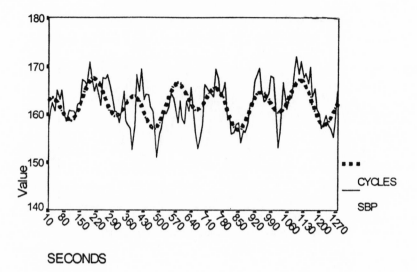

FIGURE 6.6. The combined 183- and 427-second cycles, superimposed on SBP time-series data.

structed cycles (contained in the variables "c427" and "c183") and the original time series ("sbp"); see Table 6.1. Furthermore, one can also show that if multiple regression is performed to predict "sbp" from "c427" and "c183," the amount of variance explained by these components in the multiple regression results is nonoverlapping and corresponds to the estimated percentage of variance found in the periodogram analysis.

The estimated amplitudes of the cycles can be computed from the A and B coefficients. For any given frequency, the amplitude R for the cycle = $(A^2 + B^2)^{1/2}$, where A and B are the coefficients in the Fourier transform. For the 183-second cycle, the estimated amplitude $R = (2.21451^2 + 2.39490^2)^{1/2} = 3.26$; for the 427-second cycle, the estimated amplitude $R = [(-2.175)^2 + (-.251)^2]^{1/2} = 2.19$. An important consideration is whether the amplitude of any "major" periodic components is large enough in magnitude to be of any clinical or practical significance. In this case, as we see in Figure 6.1, in the original time series, SBP is varying from about 150 to 170 mmHg, which is very substantial variability. In addition, the 3-minute cyclic variations have an estimated amplitude of about 3.26 mmHg above and below the mean, and this range of 6.52 mmHg from the lowest to the highest points of the cycle seems large enough to be notable. (It is, in fact, about half the mean size of change in SBP in many stress studies where the observed mean elevations of SBP tend to be about 9–15 mmHg; Linden, 1987.)

It is important to rule out likely artifactual explanations for such phenomena before trying to interpret them as evidence of cycles in physiology or behavior. One possible source of artifact is the occasional recalibrations that the Finapres instrument requires. It is conceivable that the low values of SBP that appear at about 3-minute intervals are an artifact of the measurement process, instead of a naturally occurring variation in blood pressure. Several kinds of evidence suggest that the 3 and 7 minute cycles are not purely artifactual. First of all, only half of the subjects in this research program (Warner & Stevens, 1991) showed these 3- to 7-minute cycles; some showed minimal variability in SBP; others showed large amounts of variablity that were not at all cyclic. Second, other researchers independently report quite similar cycles in SBP using completely different types of subjects, measurement methods, statistical analyses, and tasks (Kushner & Falkner, 1981; Benton & Yates, 1990). However, the possibility of artifact due to instrumentation should always be taken into consideration when an analyst is interpreting results of time-series studies. Additional evidence from converging methodological approaches is necessary in order to be confident that instrumentation artifacts can be ruled out as the cause of any observed patterns in time-series data.

A Suggested Index of Cyclicity Derived from the Periodogram

One index that may be reported from spectral analysis (or from periodogram analysis) is the percentage of power within just one frequency band. Another possible useful index of cyclicity is the sum of power in a set of neighboring frequency bands. An index of "cyclicity" or "rhythmicity" can be created by selecting a set of frequencies that are of interest (either a priori, because they are known to be important from past research, or post hoc, because they happen to account for a large proportion of variance in a particular case) and summing the percentage of power or variance that is accounted for by those frequencies (as in Warner et al., 1987). A significance test for a sum of power integrated across a band of frequencies is provided by Grenander and Rosenblatt (1957), and an example in which this test is applied to heart rate data for children at play is given by Wade et al. (1973).

When time-series data on many subjects are available, summary indexes such as this suggested cyclicity index can be examined (either as an independent or a dependent variable) to see if they are related to other variables (such as individual subject characteristics or manipulations of the task or situation). For instance, Wade et al. (1973) found that a higher percentage of power was contained in low frequencies (corresponding to relatively long cycles) for children playing in pairs than for children playing alone or in groups of three.

Chapter Summary

An estimated power spectrum is a smoothed version of the periodogram. There are both advantages and disadvantages to this smoothing procedure. Smoothing results in more reliable estimates of the power within each frequency band. On the other hand, when there are two or more distinct cyclic components, smoothing can make it difficult to detect them separately. Whether the researcher chooses to base interpretation or description on a smoothed spectrum or an unsmoothed periodogram, it is important to go back to the original time-series data to assess whether there really do seem to be "cycles" in it that match the periods identified by the spectrum or periodogram. Reconstructing the sinusoids from the Fourier coefficients generated in the periodogram analysis is a good way of assessing whether the patterns suggested by the periodogram are a good description of the behavior of the time series.

For some purposes, use of the periodogram may be preferable to use of the spectrum. The periodogram is more easily interpreted in terms of partitioning of variance, and it is typically easier to distinguish contributions of neighboring frequencies in a periodogram. The Fisher test of periodogram ordinates is conservative, in that it controls for the inflated risk of Type I error that occurs when the largest peak in a periodogram is examined without any a priori prediction.

When many time series are available for periodogram analysis, the deficiencies of the periodogram may be less cause for concern. If the pattern seen in one periodogram replicates across many other periodograms, either within the researcher's own data or in past research, then the researcher may be more confident that the pattern description provided by the periodogram is reasonably apt. Furthermore, if the cycles suggested by the periogram analysis do seem to provide a good fit to the original time-series data, when graphs similar to those in Figures 6.4, 6.5, and 6.6 are generated, this also increases the researcher's confidence that the periodogram analysis provides a reasonably good description of the pattern in the time series.

The advantage of the power spectrum is that the smoothed estimates of power show less sampling error than do the unsmoothed estimates of intensity at each frequency. If only one time series is available for analysis, then it may be more conservative to use spectral analysis and to avoid making the kind of claims of relatively more precise estimation of cycle lengths that are sometimes made after examination of a large number of periodograms.

Next, Chapter 7 summarizes the methods of analysis that are used to assess cycles in univariate time series (from Chapters 4 through 6) and makes some suggestions about ways of combining results of these analyses when many time series are available. Suggestions for reporting univariate periodogram or spectral analysis results will also be provided in the next chapter.

Summary of Issues for Univariate Time-Series Data

Introduction

This chapter summarizes the chapters that presented methods for the analysis of trends and cycles in a single time series (Chapters 3 through 6). The following issues are considered in this chapter. First, recommendations are made about the information that should be reported in the "Results" section of a paper that presents a description of patterns in a single time series. Second, there is a discussion of research involving the collection of time-series data on many individual subjects, considering ways of summarizing the descriptions of pattern across many time series. Finally, suggestions are made about ways to deal with nonstationarity in time series, that is, situations in which the parameters of the cycles (either their cycle length or their amplitude or phase) are changing over time.

The previous chapters outlined several methods for the description of pattern in time-series data. Chapter 3 presented methods for modeling linear or nonlinear trend using ordinary least squares (OLS) regression. Chapter 4 introduced harmonic analysis, which uses regression to fit a sinusoid of known period or cycle length to a time series. Chapter 5 presented the periodogram, which can be used to partition the variance of a time series into variance accountable for by each of $N/2$ periodic or cyclic components; periodogram results can be used to decide whether there are one or more major periodic components in a time series (which might then be modeled using the harmonic analysis methods from Chapter 4). Chapter 6 described how the periodogram can be smoothed or modified to yield a more reliable estimate of power across frequencies called a power spectrum; examination of the power spectrum can be another means of deciding whether there are major periodic components in the time series that may be modeled using harmonic analysis.

Suggested Guidelines for Analysis of Reporting Data from a Single Time Series

A researcher who has just one time-series data set may use one or several of the methods from Chapters 3–6 to describe a pattern. A typical sequence of analyses for a single time series would be the following:

1. Preliminary data screening, to assess whether the data are appropriate for periodogram or spectral analysis—Chapters 2 and 3 described methods for preliminary data screening.

2. Assessment of a linear or nonlinear trend in the time series, using OLS regression to model the trend and assess goodness of fit—If there is a significant trend, trend components should be removed before attempting to detect cycles in the data by using periodogram or spectral analysis.

3. If the cycle length of major periodic components is already known, then the investigator may choose not to do periodogram or spectral analysis but to go directly to the next stage: harmonic analysis. However, if the cycle lengths of periodic components are *not* known a priori, then periodogram analysis and/or spectral analysis are applied to the time-series residuals (after removal of any trend components) to identify major periodic components. Based upon examination of the periogram or the power spectrum, the data analyst decides whether there are one or more (perhaps several) cyclic components in the time series. If there are, a preliminary and approximate estimate of cycle length(s) can be obtained by looking at the periodogram or power spectrum. This may be the set of results that is reported, without any additional follow-up using harmonic analysis. Alternatively, harmonic analysis can be used as a follow-up, both to try to make the estimate of the cycle length more precise and to produce graphical illustrations of the fitted cyclic components.

4. Once the cycle length is at least approximately known, either from past research experience or from the results of a periodogram or harmonic analysis, harmonic analysis (Chapter 4) can be applied to the residuals from trend removal. The cycle length can be varied, in the neighborhood of the initial estimated cycle length using "grid search" methods until a best-fitting cycle length is found. A graphical illustration that shows the fitted sinusoid plotted with the original time-series data (as in Figure 4.1) can be an excellent means of making it clear what features of the time-series data are being modeled.

There are, unfortunately, few guidelines for writing up or presenting the results of harmonic, periodogram, or spectral analysis. Because some computer programs do not provide any significance tests, many past journal article "Results" sections have not included any judgments about significance. This chapter presents suggestions about the things that might well be included to make a "Results" section complete and clear. A complete "Results" section for a univariate spectral analysis ought to include the following:

1. The "Results" section must include a clear description of the X_t time-series data, including the sampling frequency (Δt) and the length of the time series (N). The author should make it clear what rationale was used to determine the sampling frequency and data record length. It should be stated explicitly what cycle lengths, if any, were predicted. It may be helpful to remind readers who are unfamiliar with spectral analysis that the frequencies included in the periodogram or spectral analysis are based on N and Δt, and to explain how these terms determine the range of frequencies that can be detected in the study. (Alternatively, this information could be presented earlier, in the "Methods" section.)

2. Comments on preliminary data screening should be included: Were scores on the time-series variable approximately normally distributed? Do the mean, variance, range, minimum, and maximum values appear reasonable? If there is a problem with the distribution shape of the time-series variables such as outliers, explain what data transformations (such as log) are applied to the data as remedies.

3. There should be a description of any trend analysis that was performed to identify and remove a linear or curvilinear trend from the X_t time series prior to doing periodogram or spectral analysis. It should be made clear whether all subsequent analyses are based on the raw data or on residuals from trend removal, which is the predominant practice. The percentage of variance due to the trend in the time-series data should be reported.

4. There should be a clear specification of the set of frequencies that are included in the periodogram or spectrum. The specific set of cycle lengths and/or frequencies can be listed in a table or in the text. If each observation corresponds to some nonunitary time counter (e.g., if each observation is equivalent to 10 seconds), then it is helpful to translate the frequency and period (given by SPSS in terms of the number of *observations*) into frequencies in cycles per unit of time (such as seconds) and cycle lengths in units of time. For instance, if there is a cycle 60 observations long and each observation is equivalent to 10 seconds, then the estimated cycle length is 600 seconds. Also, it should be clearly specified how the analysis was performed and which results are reported. If

the periodogram is reported, it should be made clear whether it is given as a set of sums of squares accounted for by each periodic component, as in the SPSS output, or as a set of Mean Squares (MSs), as in SAS output. If spectral analysis is reported, there should be a complete description of the smoothing technique(s) that were used to estimate the power spectrum from the periodogram.

5. Statistical significance tests may be presented. Large Sums of Squares in the periodogram, or large power values in the power spectrum, may be assessed to see whether they are statistically significant (using the Fisher test described in Chapter 5 or the confidence intervals based on equivalent degrees of freedom in Chapter 6). It should be made explicit what criteria were used to identify major periodic components: was this judgment based upon statistical significance testing or on a qualitative visual examination of the periodogram or power spectrum?

6. The results should include a list of the "major" or most prominent periodic components. For each major periodic component, it is useful to report the cycle length (in units of time), the percentage of variance in the time series that is accounted for by this periodic component (derived from the periodogram), the estimated amplitude of this cycle, and the statistical significance of this periodic component (using either the Fisher test of the periodogram from Chapter 5 or the confidence interval for the spectrum from Chapter 6 to assess significance).

7. It can also be extremely helpful to include a reconstruction of the major periodic components using the sine and cosine coefficients that correspond to the major periodic components. Harmonic analysis (Chapter 4) generates a predicted sinusoid for any specific cycle length that is of interest to the researchers. It can be quite useful to add the trend component to the fitted sinusoid obtained from harmonic analysis and to plot this overall prediction function on the same graph as the raw time-series data. This graph will give the reader a clear picture of the features of the time series that have been captured by the description of trends and cycles.

8. Discussion of the results should include a consideration of possible artifacts that would qualify or modify the interpretation. Could artifacts such as aliasing, leakage, or presence of a trend or extreme outliers be responsible for the obtained results? Could there be instrumentation artifacts or changes in behavior of measuring instruments over time that would explain any observed periodicity? Could "history," "regression toward the mean," or other extraneous variables account for any of the observed patterns in the time series?

Analysis of More Than One Time Series

Many studies involve collecting (similar or identical) time-series data on a number of subjects. For example, Larsen and Kasimatis (1990) had each of 72 subjects rate their daily mood for 84 consecutive days. Thus, they had one time series (N = 84 observations) for each of 72 individuals. How can data be summarized across numerous time series? West and Hepworth (1991) noted that it is possible either (1) to aggregate the time-series data for all subjects into one averaged time series and then do a spectral analysis on this averaged time series or (2) to conduct a spectral analysis for each individual subject and then summarize results across subjects (as discussed in the following two subsections).

Analysis of Aggregated Time Series

If all the subjects have been observed under the same circumstances, if all the time series begin at the same day or minute, and if is anticipated that all subjects will show similar cyclic patterns that start at the same time, then it makes sense to aggregate the time series into a single time series by averaging the observed scores across subjects at each time; then, the researcher can conduct a single spectral analysis to describe the behavior of this aggregated time series. For instance, one of the analyses reported by Larsen and Kasimatis (1990) was of this type. All their subjects began keeping daily mood-rating records on the same day, and they expected all subjects to show weekly mood cycles with the lowest mood on Monday. Therefore they computed an average mood across all subjects on each day (from day 1 to day 84) and did a spectral analysis to show that this aggregated time series did tend to show 7-day cycles. Examination of this spectrum for the aggregated time series allowed them to assess the typical or average amplitude due to a weekly cycle.

However, if Larsen and Kasimatis had looked at only this analysis, they would not have seen individual differences among subjects that they were able to detect by looking separately at each subject's individual spectra (as described in the next subsection). There are many circumstances in which it does not make sense to aggregate time-series data across subjects. For instance, if the time-series data include different numbers of observations, if the observations do not begin at the same point in time for all subjects, or if major individual differences are expected in the lengths or amplitudes of cycles for different subjects, then it may not be useful or informative to aggregate all the data into a single time series. Instead it may be necessary to look at the data separately for each subject, as described in the following subsection.

Analysis of Each Time Series Separately, Followed by a Summary of These Results across Subjects

Another method of summarizing data is to conduct a spectral analysis on each time series (in the case of the Larsen & Kasimatis study, this leads to 72 separate analyses). One overall averaged spectrum can be estimated by summing the power at each frequency across all 72 subjects and dividing by the number of subjects to create an averaged spectrum. Because power depends on the overall variance of each time series as well as the relative contributions of various frequencies, it may be useful to standardize each spectrum prior to averaging the spectra, that is, to convert each spectral estimate to an estimate of the percentage of variance at each frequency by dividing each spectral estimate by the sum of all the spectral estimates for the individual time series. When these "standardized" spectra are averaged together the information from each subject is given equal weight. If the raw spectra were averaged, information for subjects who had more overall variance in their time-series data would be counted more heavily.

An example of a study that used this method to summarize data across subjects was that of Warner and Stevens (1991). For each subject, time-series data were collected on blood pressure (measured once every 10 seconds). For each subject, a spectrum was estimated to see how power was distributed across a set of frequencies that correspond to cycles ranging from 1,280 to 20 seconds. Each spectrum was then divided through by the sum of the spectral estimates across all frequencies for that subject to yield an estimate of the spectral density (percentage of variance accounted for) in each frequency band. Finally, one overall average spectrum was computed and plotted to show what cyclic components were most consistently observed across all subjects; this averaged spectrum had two broad peaks at the low-frequency end of the spectrum, which corresponded to cycles about 6 and 3 minutes long.

Yet another way to summarize information from spectral analysis across a number of subjects is to derive one or more simple summary index values for each subject and then do descriptive statistics on that index variable (or other analyses using that variable, such as correlations or tests of differences in means across groups). One possible summary index (suggested by Warner et al., 1987), is an index of cyclicity or rhythmicity. In this particular application, the five largest periodogram intensity estimates for each subject were identified (any number of periodic components could be included in such an index). The percentage of variance that each periodic component accounted for was calculated by dividing each of these intensity estimates by the sum of all the periodogram intensity estimates for that subject. Then these percentages are summed to yield an estimate of the percentage of variance due to the five major periodic components for that individual subject. This "rhythmicity" index is one of many possible ways of summarizing how much an individual's activity tends to be accounted for by a relatively small number of cyclic components. This is similar to an approach suggested earlier by Wade et al. (1973): summing the amount of variance contained in the low-frequency end of the spectrum for each subject and then converting it to a percentage of variance. "Rhythm" indexes similar to these have been used in a number of studies. Lester et al. (1985) examined the mean percentage of power in certain low-frequency bands for individual infants who were either full term or preterm; they found that the activity of preterm infants was less rhythmic than that of full-term infants.

A slightly different "rhythm" index was used by Larsen and Kasimatis (1990) to describe the tendency for each subject to show regular 7-day cycles in mood. These authors expected 7-day cycles in mood, so their "rhythm" index was the percentage of variance explained by 7-day cycles for each subject. They showed that this percentage of variance tended to be lower for the extroverted group of subjects than for the introverts. (They noted correctly that this was not because the introverts had a large amplitude in their average weekly mood cycle, but rather because the extroverts had more "error" variability or volatility of mood that was not related to weekly cycles.)

Dilemma: Whether to Do a Standard Analysis across all Subjects, or to Tailor the Analysis to Idiosyncracies of Each Time Series

A dilemma that often arises when looking at many time series is the following: On the one hand, it might be desirable to tailor the analysis to the characteristics of each time series; for instance, in spectral analysis, it may be desirable to trim the time-series length so that the N on which the set of fitted Fourier frequencies is based comes close to matching the cycle lengths that occur in that particular time series. The N of 84 in the Larsen and Kasimatis (1990) study was purposefully selected (because 84 is an integral multiple of 7 and the expected cycle length was 7 days). However, if they believed that some individual subjects had different length cycles in mood—such as 11 days—they might have trimmed the N in the time series to an integer multiple of this period (e.g., 77 days) in order to get a good fit and to minimize the problem of leakage mentioned in an earlier chapter.

However, if the data analyst applies different procedures (e.g., a different type of trend removal; analysis of a different set of Fourier frequencies; different amounts of smoothing) in order to tailor the analysis to each data set, then the difference in the analytic procedures used may partly explain different results for different individuals.

In practice, the best solution may be a compromise. The data analyst should certainly do spectral analyses on many individual time series before deciding on a "standard" analysis that seems appropriate for most of the time series. If it looks as if there is a model that works well for all or nearly all subjects—for example, if nearly all subjects have 7-day cycles in mood—then it seems reasonable to apply this one analysis to all cases. However, it may be of interest to note cases that have other patterns. There could be some individuals who show different cycles (perhaps 28-day cycles in mood). If a researcher is too quick to choose one "standard" analysis and to apply it to all cases, it is more likely that interesting idiosyncratic patterns observed in some individual time series will be missed.

Another possible outcome of interest is that the cyclic patterns observed in time-series data for different types of persons or dyads may show different types of cyclicity. VanLear (1991) collected time-series data on (observer-rated) levels of "openness" in conversations; for each dyad, the levels of openness were rated across many conversations. Spectral analysis of the time series obtained for each dyad indicated that the nature of cyclicity differed, depending upon the nature of the relationship in the dyad. Couples who had deteriorating relationships tended to show more marked cycles in levels of openness across a series of conversations than did couples who had stable relationships. Van Lear's results suggest that sometimes it is more important to look for differences, rather than consistencies, in the patterns of time series that are obtained from different subjects or dyads.

Describing Changes in Cycles over Time

Three methods for assessment of changes in cycle parameters across time are suggested in this section. These include one very simple method that can be carried out wih any computer program, provided that the time series has a large enough number of observations—that is, dividing the time series into several segments of equal length, carrying out periodogram or spectral analysis separately for the data within each segment of the time series, and assessing whether the results differ (e.g., in estimated cycle length or cycle amplitude). The other two methods (complex demodulation, band-pass filtering) are not available in the SPSS TRENDS time-series programs; other packages (such as BMDP) do provide programs for these procedures. These two methods are described only briefly, as they are not yet widely used in behavioral science time-series research.

Analysis of Segments of Time Series

The periods or amplitudes of cycles may possibly change over time. This is both a problem for the data analyst (because changes like this violate the stationarity assumptions described earlier and may make one overall spectral analysis performed on the entire time series somewhat misleading) and an opportunity (because the changes in the nature of the cycles may be a phenomenon of interest). Formal tests for violations of the stationarity assumption are available (e.g., Weber, Molenaar, & Van der Molen, 1992). The simplest possible method of detecting change in cyclicity over time is also a good method of checking for all sorts of violations of the stationarity assumption. The time series can be divided into segments of equal length; statistics (such as the mean, the variance, or the periodogram) can then be computed separately for each segment. If you have just one time series, you may just do a visual examination to assess if there seem to be substantial changes in any of these parameters. If you have multiple subjects, you can do a repeated-measures ANOVA (using segments as the within subjects factor) to assess whether any of these parameters are changing significantly over time.

For example, Warner (1992a) examined variations in the amount of talk in conversations 40-minutes long. For each subject, the 40 minute conversation was divided into four 10-minute segments; then a periodogram analysis was done for each separate segment, and a rhythm index (percentage of variance accounted for by the three largest periodogram components) was obtained for four segments for each subject. The hypothesis being tested was that rhythmicity or cyclicity might tend to increase during the conversation; and repeated measures ANOVA confirmed that, on average, rhythmicity was significantly higher in the fourth 10-minute segment than in earlier segments. This suggests that in future research on rhythms in conversation, it might be useful to allow an initial warm-up or adaptation period and to focus primarily on later parts of the conversation, if the researcher hopes to see cyclic patterning.

Complex Demodulation

A method of describing changes in cycle parameters over time is complex demodulation (described in Bloomfield, 1976, Chap. 6). This technique is described only briefly here, with an illustrative example. Essentially, complex demodulation involves estimating a specific parameter (e.g., for instance, the amplitude of a circadian rhythm cycle as it changes over the course of many days; see Babkoff, Caspy, Mikulincer, & Sing, 1991). This parameter may be graphed as a function of time to see how it varies over time. Looking back at the sunspot time-series data plotted in Figure 1.1, for instance, note that the height of the peaks in the sunspot cycle are larger at the beginning and end of the record, and much smaller in the middle. If the estimated amplitude of an 11-year cycle were plotted as a function of time, it would be relatively high during the first third of the record, relatively low in the middle third, and relatively high in the last third.

An application of complex demodulation to psychological data was given by Babkoff et al. (1991), who studied the effects of sleep deprivation on task performance over time. Complex demodulation is not available in the SPSS for Windows time-series programs, but it is available as a feature in the BMDP 1T program.

Band-Pass Filtering

Another method for description of changing behavior of cycles over time is band-pass filtering. This will be described only briefly; for details, see Hamming (1983). Essentially, a moving-filter window is created and applied to the time series in short sections; the "best-fitting sinusoid" is estimated for local sections of the time series. If there are more clear-cut cycles in the second half of the data record than in the first half, the band-pass filter output will show larger amplitude and more regular oscillations in the second half of the conversation, thus showing how the behavior of the cycles is changing over time. An application of band-pass filtering was reported by Warner, Waggener, and Kronauer (1983). Time-series measurements of respiration were band-pass filtered to show how the cycles in this variable changed in amplitude over time. A typical graph that showed the respiration cycles over 40 minutes for one subject suggested that these cycles only became large in amplitude after the first 20 minutes of conversation (see Figure 7.1).

Filtering is not provided by the SPSS for Windows program. Some filtering features are available in the BMDP 1T program. Complete treatments of complex demodulation and band-pass filtering will not be presented here; the interested reader is referred to more technical sources, such as Babkoff et al. (1991) and Bloomfield (1976) for a discussion of complex demodulation, and to Hamming (1983) and Warner et al. (1983) for more details on implementation of band-pass filtering. Either of these can be useful ways of describing changes in cycle parameters

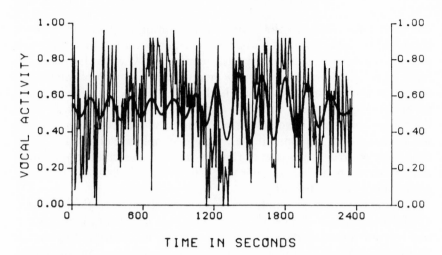

TIME IN SECONDS

FIGURE 7.1. Band-pass filter results for a time series on the amount of talk by one speaker during a converstion. The cycle length that corresponds to the center frequency for this band-pass filter was 200 seconds. Note that the cycle amplitude was small during the first half of the conversation and became much larger during the second half of the conversation. Subject 1's vocal activity (jagged line) is shown with superimposed band-pass filter output, for center frequency of $1/200$ cycles/second (smooth line). From Warner et al. (1983). Copyright 1983 by the American Physiological Society. Reprinted by permission.

(such as amplitude or phase) over time. When the time series data are clearly nonstationary, that is, when cycle characteristics are changing over time, such analyses may be necessary. A simpler alternative (described in the preceding subsection) may be to carry out the spectral analysis separately on each segment of the time series, but this is only feasible if the number of observations in the time series is reasonably large.

Empirical Example: The Larsen and Kasimatis (1990) Study of 7-Day Mood Cycles

Larsen and Kasimatis (1990) studied 7-day mood cycles, and their report of their findings illustrates many of the data analysis issues that have been discussed so far. This paper serves as an excellent model for complete and clear reporting of spectral analysis results.

Larsen and Kasimatis examined daily mood ratings for 72 students during 84 consecutive days in order to see if there are weekly (7-day) cycles in hedonic tone, or positivity of mood. In addition, subjects were identified as either introverts or extroverts based on scores on the Eysenck Personality Inventory; it was predicted that there might be mean mood differences, and possibly also differences in the amplitude of the weekly mood cycles, for introverts versus extroverts. The time-series mood data were examined in two different ways:

First, one aggregated mood time series was produced by averaging the mood of the 72 students for each of the 84 days to produce a single time series that represented the average mood of the entire set of subjects in the study over the semester. Spectral analysis was applied to this aggregated time series; the set of Fourier frequencies was chosen so that one of the components would correspond to a 7-day cycle (i.e., the N of 84 is an integral multiple of 7). A Bartlett window was used to smooth this spectrum. (No mention was made of trend removal, but visual examination of the raw time-series data in their figures did not suggest that a trend was present). The graph of this spectrum showed one very clear peak that corresponded to the predicted 7-day cycle. They used the Fourier coefficients to generate the fitted 7-day sinusoidal cycle, then graphed this predicted sinusoid superimposed on the raw data to illustrate how well it fit. The correlation between this fitted sinusoid and the raw aggregated time-series data was $r = .63$; with an r^2 of about .40, about 40% of the variance in mood was explained by 7-day cycles.

Second, data for each subject were examined separately in order to detect possible individual differences. A univariate spectral analysis was performed on the daily mood-rating data for each subject, with a Bartlett window applied to smooth the spectrum. For each subject, the spectral density estimate for the 7-day cycle was extracted. (This spectral density value is proportional to the amplitude of a 7-day cycle, or the percentage of variance explained by a 7-day cycle; it was obtained by dividing the estimated power corresponding to a 7-day cycle to the sum of the power estimates across all the periodic components, similar to the g statistic mentioned in Chapter 5, that is used in the Fisher test for significance of a periodogram.) The correlation between the extroversion score and the percentage of variance accounted for by the 7-day mood cycle was $-.32$, which suggested that extroverts tend to show less clearly cyclic moods than do introverts.

This finding was further illustrated by tabling the mean mood for each day of the week separately for the extrovert and introvert groups. Both groups tended to have relatively low or less positive moods on Monday; both groups tended to have the most positive mood scores on Friday and Saturday. However, on each weekday, the mean mood for extroverts was higher than the mean mood for introverts. The amplitude of the cycle (which is related to the range of variation) was almost identical for introverts and extroverts; that is, the mean mood on the most positive weekday was almost exactly 1 unit higher on the mood scale than the mood rating on the least positive weekday. However, extroverts showed somewhat higher variability of mood ratings within each weekday than did introverts; that is, the reason why a 7-day mood cycle fits the mood-rating data for introverts better than it does for extroverts is not because introverts show a larger amplitude cycle in mood but because extroverts show more variability in mood that is unrelated to the day of the week, perhaps due to extraneous environmental events.

The overall picture that emerges is that both introverts and extroverts tend to show a 7-day cycle in mood, with the most pleasant moods on Fridays and Saturdays; the overall mean level for the groups differs (with overall mood ratings more positive for extroverts); both groups show similar amplitudes (i.e., they feel about 1 point "better" in mood on Fridays and Saturdays than they do on Mondays; but the percentage of variance explained by a 7-day cycle is higher for introverts, while extroverts tend to have somewhat more variance in mood that is not predictable from day of the week. This paper is an excellent model for the reporting of spectral analysis results for multiple subjects.

Chapter Summary

This chapter describes how the techniques introduced in Chapters 3–6 of this book can be used in combination to describe cyclic patterns in time-series data. Suggested guidelines for reporting results were given; the Larsen and Kasimatis (1990) study provides an excellent model for thorough and clear reporting of many univariate time series.

Other sections of this chapter made recommendations about the ways in which researchers might combine and summarize information when they have time-series data on many subjects. A brief and nontechnical introduction was provided on several ways to examine changes in the parameters of cycles over time, for instance, changes in amplitude or period.

In many research situations, however, a researcher has concurrent time-series data on two different variables for the same subject or dyad. For instance, in Warner, Waggener et al.(1983), time-series data were collected on two variables for pairs of subjects. For each member of a conversation dyad, respiration and the amount of talk were measured every 10 seconds during a 40-minute conversations. The purpose of this study was to assess how these variables were related over time, both within a subject and between the subjects in the dyad. Possible questions include the following: As a speaker varies between times when she is talking more and times when she is talking less, how does her respiration pattern change? Is there a time lag such that talking has a delayed impact on respiration? Are the 3- to 6-minutes long cycles in the amount of talk that have been seen in past research accompanied by 3- to 6-minute cycles in respiration? Are cycles in talk and respiration synchronized between conversation partners; and, if so, what is the phase relation between them? These questions can be addressed by looking at bivariate methods for time-series analysis. Next, Chapters 8–10 outline a series of methods for examination of bivariate time-series data, beginning with relatively simple techniques and going on to consider more complicated analyses including cross-spectral analysis.

CHAPTER 8

Assessing Relationships between Two Time Series

Introduction

Previous chapters in this book discussed ways of detecting and describing a pattern within a single time series. However, in many research situations, time-series data are collected concurrently on two variables, and the goal of bivariate analysis of time series is to detect and describe any statistical relations between these two time series.

When no cycles are present, an assessment of the bivariate relation between a pair of time-series can be carried out using time series regression models (as described by Ostrom, 1978). When cycles are present, cross-spectral analysis may be useful as a means of assessing the bivariate relation between a pair of time series that takes into account the possible existence of coordinated (or uncoordinated) cycles in one or both series. There are multivariate generalizations of statistical methods for assessing relations among time series, but few of the widely available computer programs provide for three or more time series, and only the bivariate case (two time series) will be considered here. In principal, however, it is possible to ask questions such as the following: Is time series Y predictable from time series X when time series Z is statistically controlled?

Throughout this chapter, the two time series will be denoted X_t and Y_t. In a typical time-series regression, predictions of Y_t are made from previous values of X ($X_{t-1}, X_{t-2}, \ldots, X_{t-k}$), usually controlling for or partialing out previous values of Y ($Y_{t-1}, Y_{t-2}, Y_{t-k}$) as a means of removing serial dependence in the dependent variable. (Of course, in some research situations the researcher wants to look at Y as a predictor of X, in addition to X as a predictor of Y). For good and relatively brief introductions, see Gottman (1981b, Chap. 25) or Ostrom (1978). Trend or cyclic components can be added to this basic time-series regression model, although this has not often been done in past research. (Often, a time-series regression model is applied to residuals from trend and cycle removal that is performed as a preliminary and separate stage of data processing.)

When one or both of the time series X and Y include cycles, it is desirable to use bivariate methods that take these cycles into account. Cross-spectral analysis (in Chapter 9) is a generalization of univariate spectral analysis methods to the bivariate case (a pair of concurrent time series). As the next chapter will explain, a cross-spectrum can provide a great deal of information about the relation between a pair of time series, including the percentage of variance within a particular frequency band in Y that can be accounted for by that same frequency band in X, and description of lead–lag relations between time series. Sometimes cross-spectral analysis results are reported as a description of the data; however, such results can also be used to guide the researcher toward specification of a time-series regression model to describe the statistical dependence between two time series.

It can be difficult to interpret cross-spectral analysis results because of the inherent complexity of the analysis. Before introducing cross-spectral analysis, therefore, this chapter presents the components of the relationship between a pair of time series broken down into simpler features that can be assessed using much simpler statistical analyses. This preliminary presentation is motivated by two goals. First, some researchers may find that some of the simpler analytic methods presented in this chapter provide more direct and interpretable information to answer their questions about relations between time series. If this is the case, these researchers may want to report the simpler statistics in addition to, or instead of, the more complicated cross-spectral analysis results. Second, the basic vocabulary for describing relations between time series (such as lead–lag relations between X and Y) can be introduced more easily in the context of simpler analyses that examine just one aspect of the relation between time series at a time. This should provide the necessary conceptual background so that these terms will be understood when they are employed in Chapter 9 to describe the information available from certain components of the cross-spectrum.

The discussion begins with the simplest possible analysis: an unlagged correlation between the X and Y time series. This correlation can provide useful information, but it does not take into account many of the features of the data that the researcher may need to look at separately. Each section introduces an additional tool that makes it possible to obtain more detailed information about components of the correlation between X and Y—for instance, how the correlation between X and Y varies as a function of the time lag between series; also whether this correlation between series is partly due to autocorrelation within each of the time series. At each step, a simple analysis is introduced to deal with one more question arising from a consideration of a pair of time series. Many data analysts do not believe that any of the "intermediate stage analyses" introduced in this chapter are appropriate, because they do not take all the possible aspects of cross-correlation into account; they would argue that the only "correct" analysis is the last one presented, which includes prewhitening of one or both time series. However, I think that dealing with the problem of analysis by considering these intermediate steps offers better comprehension of the final, more complicated analysis; and I also believe that, in some research situations, a simpler analysis such as an unlagged Pearson r between time series may be the most informative (provided, of course, that the researcher is clear about what this analysis does and does not take into account).

The most familiar tool that a data analyst might use to see whether a pair of continuous, interval/ratio variables are linearly related is the Pearson correlation. If the time-series variable is dichotomous, this correlation would take the form of a phi coefficient; throughout this book, continuous variables are assumed. See Gottman (1981a) for a discussion of spectral analysis of dichotomous time series data. If the relation between the time-series variables X and Y is not linear, then data transformations or different statistical methods are required to detect these nonlinear relations (just as in any type of correlation or regression analysis; the log of X and log of Y may be linearly related for some types of data).

Unlagged Pearson *r* between Time Series

An obvious and simple initial approach to examining the relationship between a pair of time series (X and Y) would be to treat the two time series as two variables and compute a Pearson *r* between them. This chapter begins with this simplest possible approach and then develops more complex analyses that attempt to solve some of the problems inherent in this overly simple initial approach to analysis. Ultimately the vocabulary from this chapter (e.g., lead–lag relations) will be useful in understanding the output of a more general and more complex tool for understanding the relations between a pair of time series—cross-spectral analysis (presented in Chapter 9).

A typical research situation involves collection of time-series data on two variables, X and Y. For example X_t and Y_t might represent the percent of time spent talking by persons X and Y in each 5-second time interval, for time counter t ranging from 1 to N. A simple unlagged Pearson *r* can be calculated to assess overall covariation between these time series by treating X as an independent variable and Y as a dependent variable in a standard Pearson correlation, as shown here.

	X	Y
Time 1	X_1	Y_1
Time 2	X_2	Y_2
Time 3	X_3	Y_3
	.	.
	.	.
	.	.
Time N	X_N	Y_N

This unlagged Pearson r may be a useful result to report as an index of overall coordination in bivariate time-series research. This correlation summarizes the overall amount of covariation between the time series when trends, cycles, and random variations within each time series are all considered together. However, this simple Pearson r does not statistically control for the effects of autocorrelation within each time series. It certainly does not provide a basis for any causal inference about the relation between X and Y. It does not detect time-lagged or time-delayed relationships between the time series. The three main problems or issues in interpretation of the simple Pearson r between time series can be briefly summarized as follows (details on each issue are in the following sections of this chapter):

1. The observations within each time series are not (in general) independent of each other; for this reason, conventional statistical significance tests of the correlation are biased and inappropriate.
2. Many types of serial dependence *within* time series can lead to spurious correlations *between* time series. The possible existence of time-lagged dependency within each time series should be taken into account when the analyst is looking for relations between two time series.
3. X and Y may be most strongly correlated at some time lag; this would not be detected using a simple Pearson r.

Each of these issues will now be considered in more detail, and for each problem an analysis that addresses this problem is described.

The Problem of Nonindependence
of Time-Series Observations

Usually the observations within a time series are patterned in some way (trends, cycles, or other types of serial dependence) that create autocorrelations among observations (Kenny & Judd, 1996). Because the observations are not independent, conventional significance testing procedures cannot legitimately be used to assess the statistical significance of the Pearson r. If we want to perform significance tests, then it is necessary to model and remove the serial dependence in the data (trends, cycles, or other patterns) in order to obtain independent residuals that can be used to estimate error variance and obtain unbiased statistical significance tests. This can be done through a process called "prewhitening."

Prewhitening: Modeling and Removing Serial Dependence

When a data analyst models the serial dependence in a time series X, a fairly general form of the model to describe all the serial dependence in the X time series is as follows:

$$X_t = \mu + bt + R\cos(\omega t) + \phi_1 X_{t-1} + \phi_2 X_{t-2} + \ldots + \phi_k X_{t-k} + \epsilon_t$$

where X_t is the observed value of X at time t; μ is the overall mean of series; bt is a linear trend component; ω is the frequency of sinusoid in radians $= 2\pi/\tau$; $R\cos(\omega t)$ is a sinusoidal cycle of frequency ω, period $\tau = 1/(2\pi\omega)$, and amplitude R; and $\phi_k X_{t-k}$ is a lagged autoregressive term that models predictability of Xt from the value of X k time units earlier.

In its most general form, this model can include the following: (1) nonlinear in addition to linear trend components; (2) any number of cyclic components; and (3) any number of autoregressive components with various time lags. If we want to remove all these forms of serial dependence (trend, cycle, autoregressive serial dependence) and obtain white noise residuals, then we just subtract all these terms from both sides of the equation:

$$\epsilon_t = X_t - [\mu + bt + R\cos(\omega t) + \phi_1 X_{t-1} + \phi_2 X_{t-2} + \ldots + \phi_k X_{t-k}]$$

When the trend, cycle, and autoregression components are subtracted from a time series X_t to yield residuals ϵ_t, this operation is called "prewhitening"—it transforms the serially dependent X time series into "whitened" or white noise ϵ_t residuals. If the model describing the sources of serial dependence in the X time series is complete and correct, then the ϵ_t residuals that are obtained by the prewhitening operation will be white noise or uncorrelated. The whiteness or randomness of residuals from a model can be checked using the Box–Ljung test (see Chapter 3).

In principle, the residuals obtained by this prewhitening process provide the independent estimates of error that can be used to calculate significance tests for model parameters, such as the overall multiple $\underline{R}$ of a time series regression equation.

Prewhitening is also used when the researcher wants to isolate the moment-to-moment random variability in a time series from trend, cyclic, or other predictable components before looking at relations between time series (see the next section) in order to eliminate correlations between trends or cycles as possible spurious sources of correlations between time series.

Serial Dependence within Time Series and Spurious Correlations between Time Series

Spurious correlations can arise between two time series because they happen to share similar trends or cycles. However, as noted later in this section, not all shared trends and cycles should be dismissed as spurious; some theories of social psychological process would predict just such shared patterns over time (Chapple, 1970; Field, 1985; Warner, 1988). Historically, however, most time-series analysts have viewed trends and cycles as a potential source of artifactual or spurious correlations between time series.

The presence of a strong linear trend in time series X and a strong linear trend in time series Y will tend to produce a large correlation between the two time series (a large positive correlation if the trends are in the same direction; a large negative correlation if the trends are in opposite directions). A famous empirical example is the frustration–aggression inspired study of the relation between cotton prices and lynchings (Hovland & Sears, 1940; Hepworth & West, 1988). In general during the period that was studied, cotton prices went up over time and lynchings went down. The overall negative correlation between these variables may be due at least in part to these opposite direction trends; it would be stronger evidence that variations in cotton prices might "cause lynching" if deviations from these trends were related. This would help to rule out the alternative explanation that the only reason for the observed negative correlation between cotton prices and lynchings was the inverse relation between the rather strong trends in these two time series. If the data analyst wants to know whether cotton prices really have a direct effect on lynchings, it may be more appropriate to correlate residuals from linear or nonlinear trends in cotton prices with residuals from the linear trend in lynching frequencies. If these trend residuals are correlated, then it seems more plausible that cotton prices are really related to lynchings. (Much of the controversy about the proper way to analyze these data has involved the nature of the trend that needs to be removed from the two time series; see Hepworth & West, 1988, for a discussion.)

The presence of cycles of the same length in both time series may also create spurious correlations. For instance, if ice cream sales tend to rise during the hot summer months and if homicide rates tend to rise during the hot summer months, then these similar seasonal variations will produce a spurious correlation between time-series data on ice cream sales and time-series data on homicides.

More generally, any type of serial dependence within a time series makes it possible that observed correlations between time series are spurious or that they do not provide unbiased estimates of the real strength of relationship between time series. Trends and cycles are the simplest types of serial dependence, but there can be more subtle types of serial dependence such as autoregression.

Because various types of serial dependence (including trends, cycles, and autoregressive serial dependence) within time series can give rise to "spurious" or misleading correlations between time series, most data analysts (e.g., Gottman, 1981b; West & Hepworth, 1991) recommend prewhitening one or both time series before doing any bivariate analyses, using methods that are based on the influential work of Box and Jenkins (1970). Prewhitening was described in detail in the previous section, where it was used as a means of obtaining the independent residuals needed to set up unbiased statistical significance tests; here it has another application. Prior to looking for correlations between two time series, the data analyst typically removes serial dependence from one or sometimes both of the time series. As already noted, removing serial dependence from a time series is called prewhitening.

Prewhitening can take many forms, and it may include any combination of the following: removal of a linear or curvilinear trend; removal of cycles; fitting a Box–Jenkins ARIMA model to the time series and using the model to remove serial dependence from the data (Box & Jenkins, 1970). Prewhitening is sometimes applied only to the time series that represents the dependent variable; but more often it is applied to both time series. After removing trends, cycles, and/or other forms of serial dependence from the data, the residuals from each time series are correlated with the residuals from the other time series. If a significant correlation remains, then this is presumably not due to spurious influences such as shared trend or shared cycles. The remaining correlation, which controls for or partials out any serial dependence within the time series, may represent a direct or causal influence of one time series on the other.

However, as always when working with nonexperimental data, the data analyst should be cautious when trying to infer causal influence from correlational data. It is true that the correlated trend or cycle components of two time series could be due to the impact of some third variable on the time series. For instance, the seasonal cycles in ice cream sales and homicide rates in the earlier example are both probably due to the seasonal variation in a third variable (temperature). Therefore, it is true that removing cycles for the data rules out one plausible rival hypothesis. However, it is virtually impossible to rule out all plausible rival hypotheses or sources of spuriousness when the time-series data are purely observational rather than experimental.

It is still possible that the white noise components of two time series are driven by the same third variable; if this were the case, then a significant correlation between the prewhitened time series (i.e., the series with trends and cycles removed) would not necessarily indicate a direct or causal link. As a hypothetical example to illustrate this point: Suppose that a researcher does a psychophysics magnitude estimation experiment using a pair of subjects. The stimuli are generated randomly by a computer, and on each trial the two subjects independently judge the same stimulus. After any trends or cycles are removed from each subject's magnitude estimation data, the remaining data consist of residuals that should look like white noise. If a correlation were performed on these white noise residuals between the two subjects, there would be a high correlation between them, but this would *not* mean that one subject's judgments influenced the other subject's judgments. The correlation would arise because each subject's judgments were influenced by the same third variable, that is, the randomly varied magnitudes of the stimuli.

The point of this argument is that, even when data have been prewhitened before the analyst has looked at relations between time series, caution must be used in making causal inferences. Correlations between residuals could conceivably be due to some third variable. Unless the independent time-series variable has been experimentally manipulated and other extraneous variables have been controlled, it is extremely difficult to make causal inferences from correlational analyses of time-series data.

The argument so far suggests that correlations between the white noise residual parts of two time series *may* be spurious (although many data analysts assume that these correlations are good indicators of causal influence or, at least, better indicators of causal influence than correlations based on the overall time series). Next consider a related question: Are correlations that arise from shared trends and/or shared cycles *necessarily* spurious?

In econometric analysis, correlations between time series that are due only to shared trends are generally dismissed as spurious. Many economic indicator variables increase simply as a function of the passage of time, and so "passage of time"—or "maturation," to use the Campbell and Stanley (1966) terminology for threats to internal validity—becomes a plausible rival hypothesis to explain away many of the apparent correlations between time-series variables. Similarly, many economic indicator variables show seasonal variations that are related to changes in the weather and the occurrence of annual events such as Christmas shopping. Seasonal variations in ice cream sales and homicide rates are much more likely to be due to the effects of temperature on these behaviors than to any direct influence of ice cream on homicide.

However, many social psychological theories about relationship development focus on trends (or cycles) as meaningful and important patterns; and it is possible that the existence of shared trends (and/or coordinated cycles) between persons is in fact a major basis of coordination between them (Cappella, 1996; Chapple, 1970; Field, 1985; Warner, 1988). When the theory at stake involves the possible presence of coordinated or synchronized cycles, recommendations for analysis of bivariate time-series data should take into account this interest in cycles (as an important component of the data, not merely as a potential artifact). Most orthodox sources (e.g., Box & Jenkins, 1970; West & Hepworth, 1991) suggest that the data analyst should look only at correlations between residuals from prewhitening (after trends, cycles, and other serial dependence have been removed from one or both time series).

It can be useful to see how these residuals from prewhitening are correlated between time series, because this shows how moment-to-moment variations in behavior (that depart from predictable trends and cycles) are statistically related. However, it is also important to include shared trends and correlated cycles in the description of bivariate time series, particularly when theories of social influence include ideas such as mutual adaptation of level, or synchronization of behavior cycles over time (Cappella, 1996; Chapple, 1970; Warner, 1988). An example of mutual adaptation of level over time is this: as person X increases self-disclosures, person Y increases self disclosures; this would produce correlated trends. An example of coordinated cycles occurs in conversational vocal activity: as person X varies her amount of talk, person Y varies his amount of talk such that Y is talking least when X is talking most.

Researchers should be hesitant to make strong causal inferences based on correlations between *any* of these components of time-series data (trends, cycles, or residuals). Correlations between all three pairs of components provide useful information that can be an important part of the description of the process over time. Theory must be the researcher's guide: some theories about the development of social relationships would lead a researcher to look for correlated trends; some theories would lead the researcher to look for correlated cycles; others lead the researcher to examine the moment-to-moment variability in behavior that deviates from these trends or cycles.

Assessment of Time-Lagged Dependence between Time Series

As noted earlier, it is possible, and indeed often very likely, that person Y's response to person X's behavior will occur after a brief time lag. This time-lagged dependence would be missed if the only analysis applied to the data happened to be an unlagged correlation; this section introduces lagged correlations as a tool for describing time-lagged dependence between time series.

The researcher's ability to detect time-lagged dependence is limited by the choice of sampling frequency (see Chapter 2). For instance, if it typically takes a mother 5 seconds to respond to a change in her infant's behavior but their behaviors are sampled every 10 seconds, then within this time-sampling scheme mother and infant behaviors will appear to occur simultaneously. The 5-second time lag that may exist in the actual behavior can only be detected in the time-series data if samples are taken more rapidly than this time lag, for instance, one sample per second. If the sampling frequency is once per second, then if the mother responds to the infant's behavior 5 seconds after it happens, there should be a correlation between infant behavior at time t and mother behavior five observations later. It should be clear that if one of the researcher's primary interests is in these time lags, it is important to take observations rapidly enough to detect the shortest response latency that is expected to occur.

It is necessary to take time-lagged dependence into account when we begin to examine predictability between a pair of time series. If we are interested in making a causal inference (e.g., that person X's behavior is causally influencing person Y's behavior), then one of the conditions that must be met is that the presumed cause (person X's behavior) precedes person Y's behavior in time. In other words, person Y's behavior at time t should be correlated with person X's behavior at some previous time, such as time $t - 1$ (or time $t - 2$, $t - 3$, etc.). For instance, if a mother smiles, does her infant tend to smile 5 (or 10 or 15) seconds later? The lagged cross-correlation function (CCF) between the two time series can be examined to provide information about the predictability of person Y's behavior from one or more of person X's earlier behaviors.

Lagged predictability may be asymmetric. For instance, if the infant's behavior predicts the mother's later behavior but the mother's behavior does not predict the infant's later behavior, we have evidence that the relationship is not symmetrical. The mother appears to be changing her behavior in response to the infant's behavior, that is, she seems to be responsive to the infant. The infant may not appear to be responsive to the mother. This asymmetry could be described as a kind of dominance (Gottman & Ringland, 1981). Person Y's response at time t might be closely correlated with person X's response at time t (i.e., there could be a strong relationship between the time series at a time lag of 0 observations). However, it is possible that person X's behavior has a time-delayed impact on person Y's behavior at time $t + 1$, $t + 2$, or later times.

In fact, if we are potentially interested in causal influence, then we will typically prefer to look at time-lagged correlations rather than correlations at time lag 0, because we assume that causes must precede effects in time. (Temporal precedence is a necessary but not sufficient condition for causal inference.)

A commonly used tool for examination of time-lagged dependence between time series is the lagged CCF. For instance, suppose we want to know if person X's behavior at time t is related to person Y's behavior at time $t + 1$; we will compute the lag 1 cross-correlation by setting up as shown here.

$$X_t \qquad Y_{t+1}$$

$$X_1 \qquad Y_2$$
$$X_2 \qquad Y_3$$
$$X_3 \qquad Y_4$$
$$\cdot \qquad \cdot$$
$$\cdot \qquad \cdot$$
$$\cdot \qquad \cdot$$
$$X_{N-1} \qquad Y_N$$

The cross-correlation between a pair of time series can be calculated for either a positive time lag (i.e., Y's behavior k observations later than X's behavior, or the correlation of Y_{t+k} with X_t) or for a negative time lag (i.e., Y's behavior k observations earlier than X's behavior, or Y_{t-k} with X_t). The lag 0 cross-correlation is simply the correlation between the two raw time-series variables with a time lag of 0.

A graph of the lagged CCF can be examined to see whether there is an asymmetry in predictability (does Y's behavior tend to predict X's later behavior, or does X's behavior tend to predict Y's later behavior), and also to determine which time lags have the highest correlations, which would suggest how long the impact is delayed.

When the CCF is based on the raw time-series data (rather than on residuals from trend and cycles, i.e., on prewhitened data), the lagged cross-correlations can be criticized for all the same limitations described for the unlagged correlation in earlier sections of this chapter. Significance testing of lagged cross-correlations can be problematic, and spurious lagged cross-correlations might arise because of serial dependence within one or both time series.

For this reason, many researchers prewhiten one or both time series and then calculate the lagged CCF for the residuals. If the prewhitening has been successful in completely removing all serial dependence from the time series, then it should be possible to use the residuals to set up an unbiased error term for significance testing. Futhermore, if all serial dependence has been removed from both time series by prewhitening, the lagged CCF detects only covariation in the moment-to-moment changes in X and Y after all systematic patterns such as trends and cycles have been removed from the data.

Outline of Procedures for Bivariate Time Series

Ideally, we would like our statistical methods for bivariate time-series analysis to provide these three types of information:

1. Valid statistical significance tests (and/or estimates of effect size)
2. Ability to make causal inferences about relationships between time series (however, this will not typically be possible in nonexperimental research situations)
3. A relatively simple, clear, and reasonably complete description of interrelated patterns in the pair of time series

The predominant view has been that, in order to have valid statistical significance tests and in order to be able to draw causal inferences about the impact of one time series on the other, it is necessary to subject one or both time series to prewhitening (to remove trends, cycles, and any other forms of serial depedence) before looking at any relationships between the time series. It is true that in order to perform unbiased significance tests, white noise residuals are needed for the computation of error terms—and these can be obtained by prewhitening the series. However, in some research situation it may be more important to have a clear description of the patterns in the time-series data. These patterns are sometimes given little attention—sometimes even completely eliminated from the data—when researchers focus exclusively on statistical significance testing.

Furthermore, prewhitening the time series before looking at the correlation between the residuals does not necessarily provide a strong basis for causal inferences; the possible influence of shared trends and cycles is ruled out, but this does not rule out the possibility that moment-to-moment, noncyclic variability in each of the two time series could both be due to some common environmental variable that would lead to the appearance of a spurious correlation between time series even when there is no direct causal influence between the time series.

In fact, useful and important information about the pattern in the data may be lost if the data analyst removes trends and cycles from the data and does not report any information about them in the summary description of the data. In this book, the recommended methods for data analysis and data reporting tend to place greater importance on the third goal: good description of the patterns in the time series. Therefore the approach to bivariate time-series analysis that is recommended here involves retaining information about trends and cycles, and conducting a variety of follow-up analyses to assess the nature of the shared patterns between time series. The summary outline of suggested data analysis for bivariate time series that concludes this chapter explicitly focuses on variance partitioning and pattern description rather than on significance testing and causal inference. Prewhitening is often used so that the researcher can carry out significance tests and make causal inferences, but if the patterns that are removed by prewhitening are merely ignored and not included as part of the reported results, then crucial information about the pattern is lost.

Relating All the Components of a Pair of Time Series: Trends, Cycles, and Residuals

The "orthodox" recommendation for analysis of bivariate time series has been the following: remove patterns such as trends, cycles, and autoregressive serial dependence from one or both time series; then apply a lagged CCF (or cross-spectral analysis, coming up in Chapter 9) to the residuals that remain after this prewhitening operation. Performing this prewhitening has two potential advantages, as noted above: it generates the independent residuals needed for unbiased significance tests, and it may provide a stronger case for making causal inferences between time series. However, there is often something lost by carrying out this orthodox prewhitening: namely, all the information about the patterns in the data (such as cycles) that may have been of interest to the researcher in the first place.

The overall variance in a time series can be partitioned into a number of different components, as shown in Figure 8.1: trend, one or more cyclic components, and white noise. (At this point it is assumed that no additional autoregressive terms are needed to make the residuals pure

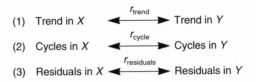

(1) Trend in X ←——r_{trend}——→ Trend in Y

(2) Cycles in X ←——r_{cycle}——→ Cycles in Y

(3) Residuals in X ←——$r_{residuals}$——→ Residuals in Y

FIGURE 8.1. Relations between each of the three components (trend, cycles, residuals) of two time series (X, Y). Note that for linear trends in X and Y, r_{trend} is artifactually either +1 or −1.

white noise once the cycles and trends have been extracted, but in practice some data are very difficult to whiten and may require a more complicated model.)

Once each time series is thought of as having these three separate and mutually orthogonal components—first, linear or curvilinear trend; second, one or more cyclic components, and third, white noise—the next logical step is to ask how each of these components in the X time series is related to the corresponding component in the Y time series. That is, how is the trend in X related to the trend in Y? How are the cycles in X related to the cycles in Y? In Figure 8.1, each double-headed arrow represents a correlation between a pair of corresponding components. (Note that a lagged correlation may be necessary to capture the relation between time series in some cases.)

The general consensus among time-series experts (e.g., West & Hepworth, 1991), based on the influential work of Box and Jenkins (1970), has been that only the third type of correlation (between the residuals from trends and cycles in time series X and the residuals from trends and cycles in time series Y) can legitimately be used in assessing theory-relevant questions about relations between trends or cycles.

Some of the implicit assumptions that are made in this typical "orthodox" approach to the analysis of time series (inspired by the econometric approach to time series) may not be appropriate when we turn to the analysis of behavioral or social interaction time-series data. First, shared trends or cycles may not be merely artifactual in social interaction; these may be the basis for the coordination of the behavior. If we remove and ignore trends and cycles, we may be removing the most important part of the information.

For instance, consider time-series data on self-disclosure of two persons. It could be that as person X gradually increases her amount of self-disclosure, person Y gradually makes a corresponding increase to match or mimic person X's level of self-disclosure. Such a process would produce an increasing trend in each of the two time series. Or consider conversational activity during the course of a day, which according to Hayes & Cobb (1979) tends to vary in about 90–100 minute cycles. When X is most talkative, Y will also be most talkative; when X is at a less-talkative point in this cycle, Y's conversational activity will also be less.

Furthermore, the lagged correlations between residuals may not necessarily detect causal connections between the time series, as some investigators seem to have assumed. Random variations in time series X and similarly patterned random variations in time series Y could be due to some randomly varying environmental variable that affects the behavior of persons X and Y identically, instead of being due to some direct causal impact of person X on Y (or Y on X).

In this chapter, it is suggested that both lagged and unlagged correlations between components of each time series (e.g., cycles in X with cycles in Y; and residuals in X with residuals in Y—as shown in Figure 8.1) are potentially useful kinds of information about relationships between the time series. However, correlations between the residuals in X and residuals in Y are not necessarily indicative of *causal* influences; and these residuals should not be the exclusive focus of data analysts.

Shared trends as a type of relationship between time series would suggest that there may be some kind of mutual adaptation between partners (or time-series variables) over time (as suggested by Cappella, 1996). However, shared trends could also very easily be spurious, that is, the trend in each time series might appear due to some environmental influence, or some intrasubject source of response pattern over time such as fatigue that happens to influence both persons in the same way. (Note that the correlation between two linear trends is artifactually either +1 or −1, so correlation is not useful information about shared trends.)

Coordinated cycles suggest the possible existence of synchronized rhythms. However, shared cycles might be "spurious"; that is, they might not be due to any direct influence of one partner on the other, but rather to the fact that both persons' behaviors are influenced by the same cyclical environment. Lagged CCFs provide information about the relative phase of cycles between time series.

Research on synchrony of menstrual cycles between women (e.g., McClintock, 1971) provides a good example of the difficulty of controlling for the shared environment (and ruling this out as a factor that might create a spurious relationship between menstrual cycles). Numerous studies report that women who live together or are close friends (and who are not taking oral contraceptives or involved in sexual relationships) may synchronize their menstrual cycles. One possible explanation for this menstrual synchrony is that women who live together are exposed to mostly the same periodic environmental stimuli (particularly the 24-hour light–dark cycles). The timing of their menstrual cycles might be due to this shared environment and not to any direct influence between the women. To demonstrate that there is some direct influence it is necessary to look at various additional kinds of data. One theory that suggests a possible mechanism for a direct relationship between menstrual cycles of females involves olfactory cues; experimentally presenting samples of odorants taken from different points in the female menstrual cycle can alter the cycle of a subject to "synchronize." Statistical and experimental controls can also be used to assess whether, after the effects of a shared environment are partialed out, there is still an association between women's menstrual cycles.

The problem arises when we try to move from mere description (the observation that menstrual cycles are synchronized) to causal inference (inferring that there is some direct entrainment of biological rhythms between women and not just a "spurious" correlation due to the shared environment). If a researcher is interested only in describing the extent of covariation between two women's menstrual cycles, a simple correlation (or lagged correlation) between time series may be a sufficient index. However, when the researcher wants to be able to make causal inferences, experimental designs that involve experimental interventions and observations under carefully controlled environmental conditions are required.

Covariation of residuals between time series suggests that moment-to-moment variations in each person's behavior (departures from trends or cycles) are predictable from these same moment-to-moment variations in the partners' behavior. This might possibly be due to a sort of "interactive repair," as suggested by Tronick (1986).

Changes in Focus of Time-Series Research

Early literature that reported univariate and bivariate time-series analyses of social interaction data were largely concerned with these basic problems:

1. Demonstrating that there was *statistically significant* patterning (departure from randomness) in individual time series
2. Deciding what types of statistical models best account for serial patterns in the data
3. Assessing whether there was statistically significant evidence that pairs of time series are "causally" related

A substantial amount of research has been done on these issues (for reviews, see Cappella, 1981; Warner, 1991, 1992c). For a wide variety of kinds of behavioral time series data, investigators have tended to find the following:

1. Most behavioral and physiological time series are nonrandom; that is, they are not white noise. There is statistically significant serial dependence in many of the behavioral time series that have been examined.

2. Most of the models that have been fitted to time-series data to account for serial dependence have been relatively simple. The choice of type of analysis often depends on the level of measurement of the time-series data. With categorical time-series data, investigators have often used Markov analysis and lagged conditional probabilities, and later, log-linear models for sequential analysis. With continuous time-series variables, investigators have used Box–Jenkins ARIMA models, lagged auto-correlations, and time-series regression. (Mathematically, a first-order autoregressive process is equivalent to a first-order Markov process; the details of the analysis differ, but the underlying process of serial dependence is the same.) In many areas of research, relatively low-order serial dependence models have been adequate to account for most of the observed serial dependence in the data (Warner, 1992c). Similarly, time-series regression models rarely include more than two or three lagged partner influence terms (and in many cases one lagged partner influence term is sufficient). A few analysts have used other, more complex models (e.g., moving average, or larger numbers of lags), but the consensus seems to be increasingly that simple models (that include a small number of autoregressive and partner influence terms) are adequate in the majority of cases.

3. When "causal modeling" approaches to time series have been employed, many researchers have followed the econometric practice of removing trends, cycles, and sometimes other types of serial dependence from one or both time series before regressing one time series on the other. This rules out the possibility of certain kinds of "spuriousness," and makes a stronger (but not irrefutable) case for causal inference than can be made from simple correlations between the time series. Again, the consensus seems to be that even when all these types of serial dependence are statistically controlled, there is still statistically significant evidence that the time series are related; that is, even apart from any existing shared trends or cycles, behavioral time-series data very often show significant evidence of relationships between partners.

These three conclusions appear to hold for a wide variety of behavior (on–off vocal activity; gaze; coded communication acts); for a wide range of kinds of subjects (newborns through adults, strangers and intimate dyads); and for a wide variety of methodologies (different sampling frequencies, data record lengths, tasks or situations) (Cappella, 1981; Warner, 1992b).

Because a reasonable level of consensus has been reached about these three initial questions, researchers are now considering a new set of questions, such as the following:

1. What *meaningful* and *simple* summary indexes can we derive from these analyses to tell us what different kinds of coordination are present between partners and how strong they are for each individual dyad we study? Many of the theories that have developed about social interaction suggest that temporal coordination of behavior is a quantitative way of assessing a more global, qualitative aspect of behavior that we might call responsiveness or rapport (cf. Field, 1985). If this is the guiding theory, then one goal in developing statistical indexes of coordination should be to identify the statistical patterns that are most closely related to observer or participant ratings of responsiveness, rapport, or other important qualitative aspects of social interaction.

Another consideration is that, other factors being equal, it is desirable to have the simplest possible data analysis. Complicated analyses have some obvious drawbacks. First of all, they may be time consuming and expensive. If the investigator has to carry out elaborate time-series analyses separately on each of 100 dyads, even with the use of computers, this can take a considerable amount of his or her time. Second, the more complex the analysis, the more difficult it is for users to understand fully what they are doing and to communicate clearly what their results mean. Third, there are at least some cases where the information that is obtained from a computationally complex and cumbersome analysis is mostly redundant with information that could be obtained fr. For all these reasons, I recommend the simplest analysis that is consistent with answering the research question.

2. How are these indexes of coordination or covariation related to external variables? The early research primarily and appropriately focused on showing that simple time-series models provide a significant fit to the data and that the relations between time series cannot easily be dismissed as spurious. Furthermore researchers also tried out a variety of time-series models, ranging from simple to complex, to assess what kinds of models would be most useful and appropriate. When these were the main research goals, it was appropriate for investigators to be concerned about valid statistical significance testing and ruling out spuriousness.

There are still useful contributions to be made in basic model identification and significance testing, particularly if new kinds of behavioral time series are assessed to see if they are well described by the models that have worked for other social behavioral variables so far. However, researchers are now in a position to start looking at new questions—how indexes derived from time-series analyses of social interaction relate to outside variables, either manipulated factors (such as the type of dyad or task) or measured outcomes (such as interpersonal attraction). In this research context, it is more important to use informative descriptive statistics than to focus exclusively on statistics that are best suited to statistical significance testing and causal modeling.

Empirical Example of Bivariate Time-Series Study

As an empirical example of a bivariate time-series problem, consider a study by Hammond et al. (1995). Repeated measurements of systolic blood pressure (SBP) were collected over time for subjects who were engaged in a psychophysical threshold detection task (critical flicker fusion sensitivity). Previous experimental research suggested that manipulations that increase blood pressure in animals (using drugs) made the animals less responsive to noxious stimuli. The goal of this study was to assess whether naturally occurring variations in SBP (either trend, cycles, or random) are related to naturally occurring variations in flicker sensitivity. Some subjects did not show any naturally occurring blood pressure variability during the testing session, and for these subjects sensitivity to flicker did not vary. Most of the remaining subjects, who did show naturally occurring variability in blood pressure, showed clear trends such that as blood pressure rose during the test session, sensitivity to flicker went down. (One subject, whose blood pressure went down during the test session, showed an increase in flicker sensitivity; thus, for all subjects who showed any substantial amount of naturally occurring change in SBP over time, flicker sensitivity was negatively correlated with blood pressure, as predicted.) It was hoped that cycles in blood pressure would be accompanied by cycles in flicker sensitivity, and the data did show some indications of cycles on the order of 3–6 minutes long in both of these variables, but the amplitudes of the cycles were very small relative to the magnitudes of the trends, and probably because relatively little of the variance in blood pressure was cyclic there was no correlation between cycles in blood pressure and flicker sensitivity.

Guidelines for Reporting Bivariate Time Series

A complete report of the findings from a bivariate time-series analysis should include some of the components listed below (depending upon the nature of the research question, it may be possible to omit some of these suggested components):

1. *Conduct a univariate analysis of both time series, X and Y.* For time series X this could include the proportion of variance and statistical significance associated with trend, cycles, and residuals. In addition, comment on the nature of any significant trends (linear or nonlinear; increasing or decreasing) and cycles (period length, amplitude, and whether the amplitude seems to be consistent across time). A graph showing the raw time-series data with a superimposed line showing the fitted trends and cycles is a good visual summary. For time-series Y the same univariate time-series results should be reported.

2. *Report on shared trends between the X and Y time series, if there is reason to believe these are not merely artifactual.* Evaluate possible existence of shared or opposite trends: if both X and Y have trends that account for large percentages of the variance, comment on this; indicate whether slopes are in the same or opposite direction. Note that it is not useful to correlate the X linear trend component with the Y linear trend component, since this correlation would necessarily be either +1 or −1; only the sign of the correlation (and not its magnitude) would be informative.

3. *Report the presence of coordinated cycles between the X and Y time series* (if there is reason to believe these are not artifactual). If both X and Y have one or a few major periodic components with the same or close to the same cycle length, the lagged cross-correlation function can be used to assess predictability between each pair of cycles. The lengths of cycles in X and Y can each by identified by periodogram or spectral analysis, then reproduced by harmonic analysis. Plotting the sinusoids that were obtained from harmonic analysis and superimposing these cycles from the X and Y series on one graph can provide a clear picture of how cycles in X are related to cycles in Y. For example, visual examination of this graph indicates whether X and Y had similar amplitudes in their cycles or different ones. It can indicate whether cycles in X have a consistent phase relationship or a lead–lag relationship with cycles in Y [i.e., do the peaks in the X cycles coincide with the peaks in the Y cycles (0 phase)]. Do the peaks in the X cycles follow the peaks in the Y cycles by some time lag or fraction of a cycle? Note that, as in the case of linear trend, if you correlate a perfectly sinusoidal cycle 20 observations long in X with a sinusoidal cycle 20 observations long in Y, there will be a correlation of +1 between these cycles at whatever time lag represents the phase relationship, so the size of this lagged correlation is not useful information about strength of relation between time series. It is more useful to ask how large are the amplitudes of cycles in X (e.g., in time-series data on

the amount of talk), and how large are the amplitudes of cycles of the same length in time series Y (volume of respiration), and what time lag separates the peaks?

An empirical example that illustrates the occurrence of coordinated cycles between two time series was reported by Warner et al. (1983). For one speaker in a conversation, the percentage of time spent talking in each 10-second time interval was the X time series variable; ventilation rate, in liters per minute, was the Y time series variable. As this speaker varied her amount of talk in approximately 3-minute cycles (with the amount of talk ranging from about .30 to .70 of the available time), her respiration volume also varied in about 3-minute cycles, with the minimum volume about 6 liters/minute and the maximum volume about 13 liters/minute.

The band-pass filter outputs that illustrate these cyclic variations in amount of talk are graphed in Figure 8.2, reprinted from Warner et al. (1983). The solid line represents cyclic variations in the percentage of time spent talking; the dotted line represents cyclic variations in ventilation rate in liters per minute. The peaks in the respiration cycles lagged just a few seconds behind the peaks in the cycles in the amount of talk,

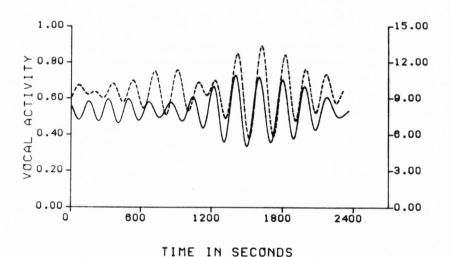

TIME IN SECONDS

FIGURE 8.2. Band-pass filter results illustrating relations between 200-second-long cycles in amount of talk and ventilation rate for a speaker during a 2,400-second-long conversation. Subject 1's vocal activity (solid line, scale on left Y-axis) and ventilation during speech (dashed line, scale on right Y-axis), both filtered at $\frac{1}{200}$ cycles/second. From Warner et al. (1983). Copyright 1983 by the American Physiological Society. Reprinted by permission.

suggesting the the maximum impact of talk on respiration was delayed a few seconds.

Another thing that can be seen from Figure 8.2 is that this coordinated cycling between respiration and the amount of talk only occurred in a strong and regular pattern from the 1,200-second time market to the 2,400 time marker; during the first 1,200 seconds of the conversation, these patterns were less cyclic and less coordinated. Large amplitude and closely coordinated cycles only appeared during the last 1,200 seconds of the conversation. Thus, it may have required some period of adjustment for this coordination between this speaker's vocal activity and ventilation to emerge.

To summarize—looking at the lagged CCF and at graphs of the fitted sinusoids (or band-pass filter outputs) that represent cyclic components of the X and Y time series can answer questions such as the following: Do peaks in time series X correspond to peaks or troughs in time series Y? Do cycles in Y lag (or lead) cycles in X by a relatively constant time lag? Chapter 12 discusses factors that might produce cycles in behavior and physiology, and mechanisms through which behavioral and physiological cycles might sometimes become coordinated.

4. *Report whether the residuals (that represent moment-to-moment variations away from trends and cycles) are correlated between the X and Y time series*, if there is reason to think these covariations are not artifactual.

The lagged CCF and/or the cross-spectrum may be used to see if there are relationships between these moment-to-moment "random" variations in the behaviors of X and Y. Typically we are most interested to know the answer to this question: At what time lag(s) are the cross-correlations largest (in absolute magnitude)? For instance, if there is a lagged cross-correlation of .70 between time series X and time series Y at a lag of two observations (Y lagged two observations relative to X), and if this is a statistically significant correlation, then we might tentatively conclude that X seems to be influencing Y's behavior two observations later (and/or that Y is responding to X's behavior two observations earlier). The lagged r^2 of .49 would indicate what percentage of the variance in the Y residuals is predictable from the X residuals at the time lag of two observations. Note that if the Y residuals only accounted for .30 of the variance in the overall Y time series, then if we want to know what percentage of variance in the overall Y time series is predictable from X, we would have to multiply .30 · .49 to obtain an estimate of .147 for this proportion of explained variance. If there is an asymmetry (i.e., if we can predict Y's behavior from X's behavior two observations earlier, but there is not a significant lagged cross-correlation between X's behavior and Y's behavior two observations earlier), then we have evidence for asymmetry in social influence: X is influencing Y (or Y is responsive to X).

Asymmetry of prediction between time series does not always occur; and in fact sometimes the highest cross-correlations between time series occur at time lag 0. When this occurs, there are at least two possible interpretations: One is that the behaviors of X and Y really do covary "instantaneously." The other explanation, which most investigators would probably prefer, is that the sampling frequency was too slow to detect the time lag between X's action and Y's response. If we sample behaviors every 5 seconds but there is a 1-second time lag between X's and Y's behaviors, then our 5-second chunked data will make it appear that their behaviors were related concurrently. We would have to sample at least once per second (or faster) to detect a real time lag of 1 second. For this reason, we need to choose our sampling frequency so that it is fast enough to detect the shortest time delay in reactions that we would be interested in knowing. Certainly we cannot detect asymmetries in prediction between X and Y if the sampling frequency is too slow to detect the time lag between X's behavior and Y's response.

Conceptually, these three types of statistical interdependence (correlated trends, coordinated cycles, correlated residuals) between time series are logically distinguishable (cf. Cappella, 1996). It is conceivable that a dyad could have one, or all, or none of these types of covariation between their behaviors. Each of these types of covariation could possibly be spurious due to the influence of some third variable that affects the behaviors of both X and Y, although this seems more likely to be a problem with trends and cycles than with residuals. It is possible, however, for both X and Y to be responding to the same white noise variable in the environment and thus generating responses that have highly correlated residuals that are not directly related to each other.

Chapter Summary

This chapter reviewed several statistics that can be used to assess relations betwee two time series. First, the unlagged Pearson r was described as a possible summary index of overall coordination. However, several limitations of this analysis were noted: it does not provide the independent error terms needed for significance testing; it does not provide a basis for making a causal inference; and it does not separate out the several components of the time series (trend, cycles, and residuals) that might each be a partial explanation for any observed overall coordination.

Subsequent sections of this chapter described ways to deal with these limitations. Prewhitening the two time series can be a useful preliminary step, for two reasons: it provides the independent residuals needed to set up unbiased significance tests; and when the residuals of the two time series are correlated with each other the resulting coordination index is now restricted to coordination of moment-to-moment variations in behavior, which may be less likely to be spuriously related than trend or cyclic components. Lagged cross-correlations were introduced as a simple method for detection of time-lagged dependence between time series. When the lagged CCF is applied to the residuals from prewhitening, the resulting analysis is free from some of the problems that were identified near the beginning of the chapter: these cross-correlations have independent residuals that can be used to do unbiased significance tests (assuming that the prewhitening was successful in removing all serial dependence); the possibly spurious relations between trends and cycles in the two time series have been removed; and the cross-correlations for the prewhitened data make it possible to make inferences about how much the Y time series changes in response to changes in the X time series, k time lags earlier, with the predictable patterns in both X and Y removed from the data.

However, researchers do not always want or need to remove trends and cycles from the data. Sometimes the trends and cycles in time series are the most interesting component of the data. For this reason, even though there are times when unbiased significance tests are not available and intepretation of results may be ambiguous, it may sometimes be very useful to look at simple unlagged correlations, or CCF's based on raw data rather than prewhitened data. When prewhitening is not applied to the data, correlations between time series provide estimates of overall coordination—combining the contributions of any trends, cycles, and residuals. In some research situations, this overall coordination may be what the researcher wants to know. It is also possible to isolate each component of the time series (fitted trend, fitted cycles, and residuals when these patterns are subtracted from the time series) and to look at these components to see how they may be related across time series (see Figure 8.2 for an example of related cycles between two time series).

Next, Chapter 9 describes cross-spectral analysis, a more complicated method for assessing relations between time series. Essentially, like the CCF analysis described in this chapter, cross-spectral analysis provides information about the strength of the relationship between time series (within each frequency band, how predictable is the Y time series from the X time series?) and about lead–lag relation or phase between time series (after a peak in the X time series, how many time units later, or what fraction of a cycle later, does a peak in Y occur?). Chapter 10 provides empirical examples that show how the methods of Chapter 8 (lagged CCF, with or without prewhitening) and Chapter 9 (cross-spectral analysis, with or without prewhitening) can be applied to bivariate time-series data.

Cross-Spectral Analysis

Introduction

Earlier chapters described the use of a univariate power spectrum to identify cyclic components in a single time series (Chapter 6), and described some time-domain statistical methods (such as the lagged cross-correlation function introduced in Chapter 8) that can be used to describe the relation between two time series. This chapter describes a generalization of spectral analysis from the univariate to the bivariate case. To perform a bivariate or cross-spectral analysis for a pair of time series (denoted X and Y), it is first necessary to obtain the univariate spectrum for each of the individual time-series variables.

The cross-spectrum is essentially the cross-product of these two smoothed univariate spectra. (An alternative way of estimating a cross-spectrum is to do a discrete Fourier transform of the lagged cross-correlation function (CCF) between the X and Y time series.) The complex numbers that result are not directly interpretable but they are converted into an estimate of coherence, and an estimate of phase, for each of the $N/2$ frequencies in the spectrum. Coherence, like an $\underline{R}^2$, indicates the percentage of shared variance between the two time series at a particular frequency. Phase, like a time lag, indicates the timing of peaks in the Y series relative to peaks in the X time series at a given frequency (however, phase is usually given in terms of fractions of a cycle, whereas time lag is usually given as the number of lagged observations). Cross-spectral analysis thus offers an alternative to the methods of assessing relations between time series that were presented in Chapter 8. Cross-spectral analysis can be a useful exploratory analysis: if the data analyst does not know a priori what frequency bands account for most of the variance in each time series and/or does not know the time lag between the two time series, cross spectral analysis provides estimates of these across all the $N/2$ frequencies. Based on the cross-spectrum, it may be possible to identify some features of the data (e.g., cycles 10 observations long with a time lag of one observation, such that the Y cycles are 1/10 of a cycle behind the cycles in the X time series) that can be modeled and described using the simpler methods of analysis that were presented in Chapters 4 and 8. Alternatively, reporting coherence from one or several frequency bands in the cross-spectrum can be a useful way of summarizing how strongly (or weakly) related the X and Y time series are across the entire set of N/2 frequencies.

Components of the Cross-Spectrum: Coherence and Phase

A cross-spectrum provides information about the relationship between a pair of time series. Just as the variance of a single time series was partitioned into a set of N/2 frequency components in univariate spectral analysis, in cross-spectral analysis we obtain information about the relationship between the pair of time series separately for each of the N/2 frequency bands.

For each of the N/2 frequency bands, we will want to know the following:

1. *What proportion of the variance in each of the two individual time series is accounted for by this particular frequency band?* This information is obtained from a univariate periodogram (or spectral) analysis of each of the two time series, just as described in Chapter 6.

2. *Within this frequency band, how highly correlated are the pair of time series?* We will obtain a coherence spectrum to answer this question. Within each frequency band, the squared coherence (like an R^2 in regression analysis) estimates the percentage of the variance in time series X that is predictable from time series Y, within this particular frequency band.

3. *Within each frequency band, what is the phase relationship or time lag between the time series?* That is, do changes in the Y time series occur at the same time as changes in the X time series, or is there a tendency for changes in Y to lag behind changes in X by some fixed amount of time (or by some fraction of a cycle)? We will obtain a phase spectrum to estimate this phase relationship between time series within each frequency band.

The following general strategy for cross-spectral analysis of a pair of time series X and Y is recommended:

1. Begin by doing a careful univariate analysis of each of the individual time series, as described in Chapters 1–7. Based on this preliminary analysis, the analyst can decide whether one or both time series include trends that should be removed prior to looking for cycles, and whether one or both time series have significant periodic components.

2. Any prewhitening that is judged to be necessary, based upon the trend and spectral analysis of each of the time series, can then be performed on each of the individual time series. In most cases, trend components are removed from each time series before doing a cross-spectrum. In some cases, an analyst might also remove or filter out cyclic components from each series before doing cross-spectral analysis. The decision as to whether to remove trends and/or cycles should be based on a consideration of whether these components of the time series might constitute the "real" relationship between series or whether relationships between these components should be dismissed as spurious. In my work, I use cross-spectral analysis to try to see how cycles are related between time series. Unlike the econometric approach, the data analysis strategy suggested here does not treat cycles as a spurious source of correlation between time series; instead, cycles are treated as components of the time series that the investigator may be most interested in. For this reason, this book does not (in general) recommend filtering out or removing cycles prior to doing a cross-spectrum. However, if the researcher believes that correlated cycles are a source of spurious relationship between series (as in many seasonal components of econometric time series), then these components should be removed before doing a cross-spectral analysis.

3. Using the residuals from any prewhitening that was judged to be necessary as the input data, one can carry out a cross-spectral analysis. Note that smoothing must be applied to the univariate spectra in order to obtain useful estimates of coherence. If the unsmoothed univariate spectra are combined to form a cross-spectrum, the squared coherence at each frequency will artifactually have a value of 1.00. The quadrature spectrum and cospectrum that are obtained as intermediate results in this analysis are not usually of much interest. The coherence spectrum and the phase spectrum that are derived from these generally provide useful information. Ideally, in addition to obtaining a plot of the squared coherence and phase by frequency, one will also obtain the actual computed values of coherence and phase for each frequency (these can be saved as new variables into the SPSS worksheet). This makes it possible to examine the values more closely at specific frequencies.

4. At this point the data analyst is likely to feel overwhelmed by the amount of information. For each frequency (or period), there will be estimates of the following: the percentage of variance accounted for by this frequency in time series X; the percentage of variance accounted for

by this frequency in time series Y; the squared coherence between X and Y at that frequency; and the phase relationship between X and Y at that frequency. There are several ways of narrowing down the amount of information that needs to be reported.

First, it is important to understand that the coherence spectrum has to be interpreted taking into account the univariate spectra, and the phase spectrum has to be interpreted taking into account the coherence spectrum. Squared coherence (like an r^2) is an estimate of the percentage of variance in X, within a certain frequency band, that is predictable from the variance in Y within that same frequency band. However, if there is a negligible amount of variance (or, equivalently, a very small amplitude of oscillations in the X time series) corresponding to the frequency that is of interest, then a large squared coherence is not especially important.

For instance, consider a hypothetical example in which X is a time series of blood pressure measurements. There might be a high percentage of variance (and therefore large-amplitude oscillations, perhaps as large as 10 or 15 mmHg, in blood pressure) at a frequency that corresponds to 180-second cycles but a very small percentage of variance (and therefore very small amplitude variations, perhaps less than 1 mmHg, in blood pressure) at a frequency that corresponds to 20-second cycles. If there is a squared coherence of .8 at the frequency that corresponds to a cycle of 180 seconds, then this means that 80% of these large-amplitude (10–15 mmHg) oscillations in blood pressure are predictable from variations in the other time series. In other words, in this case the sizes of the blood pressure changes that are being predicted are large enough to be of some practical or clinical importance.

On the other hand, if there is a squared coherence of .80 between the time series within a frequency band that corresponds to a very small part of the variance in the X time series (e.g., if cycles 20 seconds long have an amplitude of only about 1 mmHg in the X time series), then this implies that even if 80% of these oscillations in blood pressure are predictable from variations in the other time series, even if the coherence is high, the actual size of the changes in physiology that are being predicted is too small to be of much clinical or practical importance. Thus if a researcher is interested in accounting for variability in blood pressure, the researcher will most likely be interested in high coherence *only* when it is accompanied by a large amount of power at the same frequency in the univariate spectrum for blood pressure for one or both of the time series that are being analyzed.

We can therefore use the univariate spectra as a screening device to decide which (relatively few) frequency bands contain enough of the variance in one or both time series to be interesting, and then look at squared coherence only for those selected frequencies. Alternatively, if we have strong a priori reasons for being interested in a particular frequency band, we might use that a priori cycle length instead of, or in addition to, the sample spectrum as a basis for deciding which frequency bands to examine.

If we are interested in predictability between the time series over a wide band of frequencies, then it may make sense to average together the coherence estimates for a number of frequency bands. Porges et al. (1980) suggested that it would be useful to weight these estimates by the amount of variance in the univariate spectra at these included frequency bands. This "weighted" coherence seems like a very reasonable way to summarize information about coherence across a set of frequency bands.

Similarly, interpretation of the phase spectrum is highly dependent upon the values of the coherence spectrum. Unless there is reasonably high coherence between the two time series X and Y within a particular frequency band, it does not make sense to ask what the phase relationship between X and Y is at that frequency. The phase relationship between two time series can only be estimated reasonably reliably if the coherence is reasonably high (in fact, the sampling error of the phase estimate is inversely related to squared coherence; i.e., as coherence gets small, the error in estimation of phase gets larger). Therefore, we can and should limit our examination of phase to those frequencies where there is reasonably high coherence.

In SPSS (and in many other programs), phase between time series is reported in radians. One full cycle is 2π radians. In order to convert phase reported in radians to phase in terms of fractions of a cycle, just divide the phase value given by SPSS by the constant 2π (about $2 \cdot 3.1416$ . . .). Phase expressed as proportions of a cycle ranges from +.5 to 0 to −.5. If phase is 0, then the peaks in cycles in X occur at the same time as the peaks in cycles in Y. If phase is +.25, it means that cycles in X peak 1/4 cycle before cycles in Y within this frequency band. A phase of +.5 cycle is logically indistinguishable from a phase of −.5 cycle; that is, if X is one-half cycle ahead of Y and the cycles are occurring regularly, then it is also true that Y is one-half cycle ahead of X. Therefore if the phase is either $+\pi$ or $-\pi$ radians (or, equivalently, +.5 to −.5 fractions of a cycle), it simply means that X and Y tend to be in opposite phase. That is, a peak in the X time series tends to occur at the same time as a trough in the Y time series (and vice versa).

When we know how long the corresponding cycle is, we can convert this phase (in fractions of a cycle) to an estimated time lag. If the frequency band that we are interested in corresponds to a cycle length of 160 seconds and the phase relation between time series is .10 cycles, then the time lag is estimated to be 16 seconds, or 1/10 part of a 160-second cycle.

Coherence can be useful as a means of detecting time-lagged dependence between time series, although there are techniques (such as lagged CCFs) that provide similar information in a form that may be easier to understand.

Several outcomes are possible for the squared coherence spectrum. One is that coherence is uniformly quite low across all frequency bands. This suggests that the X and Y time series are not related within any frequency band or at any time lag. Another is that coherence is uniformly fairly high across all frequency bands. Yet another is that coherence is selectively high only within one or a few rather narrow frequency bands.

Under what conditions might we conclude that we have evidence of synchronized cycles between two time series X and Y? It is not sufficient to have high coherence. That conclusion may only be drawn if we see all of the following, either in the same frequency band or in closely neighboring frequency bands:

1. A high percentage of power of time series X is contained in this narrow frequency band.
2. A high percentage of the power of time series Y is contained in this same narrow frequency band (or, allowing for some estimation error, in a closely neighboring frequency band).
3. There is high coherence between X and Y in this same frequency band.
4. The phase relationship between the cycles in X and Y can then be estimated by looking at the phase for this frequency band.

Cross-spectral analysis may also be used to assess lead–lag relationships in data that have no cycles or in data from which cycles have been removed. For example, Gottman and Ringland (1981) were interested in lead–lag relations in mother–infant interaction. They wanted to know whether there is an asymmetry in predictability such that the mother's behavior tends to be predictable from the infant's behavior 5 seconds earlier (which would suggest that she is responsive to changes in the infant's behavior). On the other hand, the infant's behavior might not be predictable from the mother's behavior 5 seconds earlier if the infant has not yet become socially responsive. In this study the authors were not interested in behavior cycles but only in the possible existence of an asymmetrical lead–lag relationship. They used cross-spectral analysis as a tool for detection of a lead–lag relationship.

If there is a constant 5-second time lag (such that the mother's behavior is predictable from the infant's behavior 5 seconds earlier), then the graph of the phase spectrum should show a clear linear trend. Phase is given in terms of proportion of a cycle, so at low frequencies a 5-second time lag would be a very small proportion of the cycle length, and at higher frequencies a 5-second time lag would be a high proportion of the cycle length. If there is a clear linear decline in the phase relationship between the two time series from the low-frequency to the high-frequency end of the cross-spectrum, this is evidence for a clear and consistent time lag between the two time series. This is in fact the kind of pattern Gottman and Ringland (1981) found in their cross-spectrum.

Lagged cross-correlation functions are an alternative (and sometimes simpler) method for looking at lead–lag relationships in noncyclical time-series data, and an alternative way that Gottman and Ringland (1981) could have detected this time-lagged relation between the mother and infant would have been to look at the lagged cross-correlation function.

Computation of the Cross-Spectrum

Recall from Chapter 6 that the Fourier transform J_x of a time series X could be calculated for each frequency ω, as follows:

$$J_x(\omega) = \frac{1}{N} \sum_{t=0}^{n-1} X_t e^{-it\omega}$$

Let $J_y(\omega)$ be defined similarly as the Fourier transform of time series Y at frequency ω. Then the cross-periodogram $I_{x,y}$ between the time series X and Y can be calculated as follows (Bloomfield, 1976, p. 210):

$$I_{x,y}(\omega) = \frac{N}{2\pi} J_x(\omega) J_y(\omega)^*$$

The terms (such as $I_{x,y}$) in this equation consist of complex numbers. Generally the complex numbers are not of interest in themselves; however, the complex numbers that make up the cross-periodogram or cross-spectrum can be used to compute estimates of the coherence and the phase between the X and Y time series at each frequency.

An alternative way of representing the computation of the cross-periodogram is also given by Bloomfield (1976). The X and Y series can first be "combined" in the time domain by computing the lagged cross-covariance function; then the Fourier transform of this lagged cross-covariance function can be taken, to yield the cross-periodogram. The cross-covariance between the X and Y time series at lag r is given by the following, the sum taken over all values of t for which both t and $(t - r)$ lie in the range $0, 1, \ldots, n - 1$ (Bloomfield, 1976, p. 212):

$$c_{x,y,r} = \frac{1}{n} \sum x_t y_{t-r} \qquad |r| < n$$

Note that the covariance of X_t with Y_{t+r} is equal to the covariance of X_{t-r} with Y_t. Next we take the Fourier transform of this lagged cross-covariance function to obtain the cross-periodogram (Bloomfield, 1976, p. 212):

$$I_{x,y}(\omega) = \frac{1}{2\pi} \sum_{|r|<n} c_{x,y,r} e^{-ir\omega}$$

This is equivalent to the result we obtained from the previous equation (which gave an alternative method of computing the cross-periodogram).

The cross-spectrum is estimated by smoothing the cross-periodogram (just as the univariate spectrum for one time series was estimated by smoothing its periodogram). Any of the windows that are commonly used to smooth the univariate periodogram may also be used to smooth the cross-periodogram.

It is useful to convert the complex numbers that form the cross-periodogram into estimates of the coherence and phase at each frequency. Bloomfield (1976) used $g_{x,y}$ to represent the smoothed cross-spectrum that was obtained from $I_{x,y}$, the cross-periodogram.

The squared coherence [denoted $s_{x,y}(\omega)^2$] between time series X and Y at frequency ω is computed from the cross-spectra and the individual spectra of X and Y, as follows:

Let $g_{x,y}$ be the cross-spectrum (i.e., smoothed cross-periodogram) for series X and Y.

Let $g_{x,x}$ be the spectrum (i.e., smoothed individual periodogram) for series X.

Let $g_{y,y}$ be the spectrum (smoothed individual periodogram) for series Y.

Then the estimated coherence (which is interpreted like an $\underline{R}^2$) between series X and Y at frequency ω is estimated by

$$s_{x,y}(\omega)^2 = \frac{g_{x,y}(\omega)^2}{g_{x,x}(\omega)\, g_{y,y}(\omega)}$$

Note that this is analogous to one variant of the formula for a simple Pearson r^2; one way r^2 may be calculated is by taking the covariance between X and Y, divided by the variance of X and the variance of Y. Coherence is essentially the same ratio (of covariance to individual variances), but coherence is computed separately within each frequency band, ω. Note that most cross-spectral analysis computer programs (such as SPSS for Windows) report *squared* coherence for each frequency; but some may report the square root of this term. That is, some computer programs report $s_{x,y}(\omega)$, the coherence or magnitude of the cross-spectrum, rather than $s_{x,y}(\omega)^2$, the squared coherence.

Coherence (like an r^2) necessarily lies between 0 and 1 (provided that the weights used in smoothing the spectra were nonnegative). It has essentially the same interpretation as r^2. The squared coherence at each frequency may be interpreted as the estimated proportion of variance that is shared between the two time series within that particular frequency band. It detects statistical dependence between time series even when there is a time lag or phase delay between the series. Note that if we do not smooth the periodograms and just form a ratio of the unsmoothed cross-periodogram to the individual unsmoothed periodograms, then this ratio will be identically 1 across all frequencies. We can only estimate coherence using smoothed versions of the spectra and cross-spectra.

The phase of the cross-spectrum, $\phi_{x,y}(\omega)$, is calculated from the imaginary (Im) and real (Re) parts of the cross-spectrum, as follows:

$$\phi_{x,y}(\omega) = \arctan\left[\frac{\text{Im } g_{x,y}(\omega)}{\text{Re } g_{x,y}(\omega)}\right]$$

This is analogous to the computation of the phase for an individual time-series variable as a function of the A and B (real and imaginary coefficients) of its Fourier transform, as previously described in Chapter 4. The sampling error of this estimate of phase (ϕ) is a function of the squared coherence; the sampling error is small if squared coherence is large. When the squared coherence at a particular frequency approaches 0, the sampling error of ϕ becomes extremely large; in some sense, the phase is not defined for a frequency band when the coherence between time series is near 0 for that frequency band (see Bloomfield, 1976, p. 225).

Significance Testing for Coherence and Phase

SPSS does not provide significance tests for either the coherence or phase spectra. However, if significance tests are desired, detailed descriptions of procedures may be found in Koopmans (1995, pp. 282–287).

Koopmans provides graphs that show the upper and lower limits for the 80% and 90% confidence intervals (CIs) for coherence estimates (see Koopmans, 1995, Figures A9.1, A9.2, pp. 350–351). These upper and lower limits may also be calculated by hand, employing formulas that use the inverse hyperbolic tangent function (given by Koopmans).

Similarly, Koopmans (1995, pp. 285–286) outlines detailed procedures for setting up CIs for estimates of phase. Note that the width of the CI for phase at a specific frequency is inversely related to the coherence at that frequency; thus, as coherence approaches 0, the CI gets extremely wide and it is no longer possible to make a reasonable estimate of phase. In fact, when coherence equals 0, the phase is not defined.

Chapter Summary

Cross-spectral analysis results can be interpreted, along with univariate spectral analysis results, to make inferences about several kinds of relations between time series. If squared coherence is not significantly different from 0 across all frequencies, this suggests that the two time series are not (linearly) related at any frequency or any time lag. If there are high coherences at some frequencies, then this may indicate a time-lagged dependence between the two time series; the length of the time lag can be inferred from the phase spectrum. Whether the statistical dependence between the time series that makes the coherence high is due to coordinated cycles can only be assessed by looking at the individual spectra to see whether these show any clear evidence of cycles.

Next, Chapter 10 presents extensive empirical examples that illustrate the use of the cross-spectral analysis methods introduced in this chapter and also the simpler statistical methods for bivariate time-series analysis that were described in Chapter 8. In some situations, the simpler methods described in Chapter 8 may make it easier for a researcher to detect and describe relations between time series; however, some of the summary statistics that can be derived from cross-spectral analysis, such as a weighted coherence that summarizes predictability between a pair of time series across a set of frequency bands (Porges et al., 1980), may be useful.

CHAPTER 10

Applications of Bivariate Time-Series and Cross-Spectral Analyses

Introduction

This chapter shows how the exploratory data analysis methods for bivariate time series from Chapter 8 and the cross-spectral analysis methods described in Chapter 9 may be used together to develop a detailed description of the relation between a pair of time series. I will first describe some of the types of pattern that the data analyst might expect to see in bivariate time-series analysis; then I will present detailed examples that illustrate how the data analysis techniques from Chapters 8 and 9 can be used to describe various types of pattern in bivariate time-series research.

Possible Outcomes of Bivariate Time-Series Analysis

There are many possible outcomes when the relation between two time series is examined using the techniques described here; these include no relationship, an asymmetrical lead–lag relationship, and coordinated cycles. These three major outcomes are described in this chapter.

Null Outcome: Time Series Are Not Statistically Related

An initial assessment as to whether there is any relation between a pair of time series usually involves doing one or both of the following analyses, described in earlier chapters: a lagged cross-correlation function (CCF) and/or a coherence spectrum (part of the cross-spectrum). If none of the lagged cross correlations between the time series are large and none of the coherence estimates are large, then it seems likely that there is simply no relationship between the two time series. Keep in mind, however, that just as a Pearson r detects only linear and not curvilinear relations, the same limitations apply to cross-correlations and coherences. Application of a data transformation (such as log) to one or both time series before lag cross-correlations or cross-spectral analysis are done may make it possible to detect nonlinear relations between time series.

Detection of a Lead–Lag Relation between Time Series

In some social situations it is reasonable to expect an asymmetry in predictability between two time series. For instance, suppose X is a time-series measure of infant behavior and Y denotes time-series observations of mother behavior. The mother may be responding to infant behavior, but the infant may not be responding consistently to mother behavior. If this is the case, then the infant time series will lead and the mother time series will lag; mother behavior at time t will be predictable to some extent from infant behavior at times $t - 1$, $t - 2$, and so forth. (Actually, some research, e.g., of Jaffe, Stern, & Peery, 1973, suggests that infants are responsive to mother behavior at extremely early ages, so this kind of asymmetry may not occur often even in mother–infant research; but see Gottman & Ringland, 1981, for a study in which asymmetry of influence was found in mother–infant interactions.) When such a lead–lag relationship exists, the key questions we need to answer in the statistical analysis are the following: Which time series leads, and which lags? What is the time lag at which the changes in the second time series occur, relative to changes in the lead time series? How strong is the largest lagged cross-correlation?

We may also want to know how this lagged cross-correlation changes if we control for serial dependence in one or both time series, or if we control for other time-series variables that might be influencing both time-series. In addition, we may want to do cross-spectral analysis to see if the lead–lag relationship we have detected through these analyses can be understood as cycles in the two time series, with one time series cycling after the other.

While a lead–lag relation between the time series might suggest the possibility that X causes Y, we need to be very cautious about drawing causal inferences from any correlational analysis. If we are doing a time-series experiment in which we repeatedly manipulate X and measure Y at some time lag, with good controls for extraneous variables, then we have a stronger case for making a causal inference than in nonexperimental studies in which we simply measure X and measure Y repeatedly over time.

Some analysts argue that by statistically controlling for serial dependence within one or both of the time series, it is possible to rule out spuriousness as an explanation and to make a strong argument that X causes Y (if Y still is predictable from X, even after partialing out past values of X and/or past values of Y). A more conservative approach to causal inference is recommended here. Partialing out serial dependence in this way does not make it possible to make a strong causal inference; other, outside variables could still be influencing one or both of the time series. Statistical control is not, in general, as effective a means of justifying causal inference as is experimental control; as suggested elsewhere in this book, researchers who are really interested in making causal inferences should consider an experimental time-series design. Furthermore, in some research situations, such as social interactions, partialing out serial dependence may remove the very phenomena that we want to study. The recommendation that is made in this book is that the researcher should examine the relations between time series both with and without prewhitening to remove serial dependence within each time series. Which of these results are the more meaningful may depend upon the kinds of variable being studied, and the type of influence between time series that the researcher seeks to detect. If the main goal is description rather than causal inference, it may be more informative to look at the results obtained without doing any prewhitening. If the main goal is to make a causal inference, then the researcher may prefer to examine the lagged CCF results (or cross-specta) that are obtained after prewhitening one or both of the individual time series. However, even correlations between these prewhitened time-series data do not make a strong basis for causal inferences about the effect of one time series on the other. If causal inference is the goal, experimental methods are more appropriate than correlational methods of analysis that rely on purely observational/nonexperimental data collection.

Synchronized Cycles

Some social interactions involve the coordination of activity cycles, such as cycles in the amount of talk (Chapple, 1970; Warner, 1979). In the simplest situation, speaker X might show 3-minute cycles in the amount of talk (e.g., about 1½ minutes spent mostly talking, alternating with about 1½ minutes spent mostly listening). Speaker Y might show a vocal activity pattern that is almost a mirror image of speaker X; when speaker X is most talkative, speaker Y is least talkative, and vice versa. Univariate and cross-spectral analysis results for this situation would (ideally) yield the following: Univariate periodogram analysis (or spectrum analysis) of each speaker's time-series data of the amount of talk would show a single large peak, corresponding to a cycle about 3 minutes long in the amount of talk. At the frequency that corresponds to this 3-minute cycle, there would also be a high squared coherence between the two time series. The phase relationship between the two time series at this same frequency would be either $+\pi$ or $-\pi$ radians, indicating that speaker X's vocal activity peak occurs one-half cycle before (or equivalently, one-half cycle after) speaker Y's peaks in vocal activity.

Unfortunately, outcomes involving cycles are not always simple and clear cut. As noted, for example, in Warner and colleagues' (1983) analysis of cycles in the amount of talk and in respiration, it is possible (1) for neither time series to be cyclic; (2) for the two time series to be cyclic with synchronized cycles of the same length; (3) for one time series to have cycles in it that are not synchronized with cycles in the partner's activity; or (4) for one or both persons to have multiple cyclic components in their behavior or physiology.

Empirical Example: Lead–Lag Relationship between Talk and Heart Rate for One Speaker

The data for this example were collected on one speaker in a conversation. Once every 2 seconds, an automated vocal activity detection system reported the proportion of the time she spent talking (ranging from 0 to 1.0). Once every 2 seconds, a Finapres noninvasive cardiovascular monitor reported her heart rate. A set of $N = 240$ observations was used for this analysis; because each observation corresponds to 2 seconds, this was a 480-second-long data record. The raw time-series data for this empirical example are given in Appendix A as the file entitled "talkhr.sav."

A series of analyses of these data are presented to show what information can be obtained from each kind of analysis. A first approach to this data was to calculate the simple Pearson r (not lagged) between the time-series data on the amount of talk and heart rate; this yielded $r(238) = +.18$. This suggests that heart rate is only modestly positively correlated with the immediate amount of talk in the same 2-second observation period. However, this leaves open the possibility that there could be a stronger correlation at some time lag, perhaps because talking has a somewhat delayed impact on heart rate.

It is easy to become confused as to which variable leads (and which variable lags), so in the following example I will be very explicit about the order in which variables are named in SPSS commands and their location on the printouts and plots. The next analysis is a lagged CCF, with the amount of talk ("ftalk" or "FTALK") as the first variable named in the CCF command (because it is probably the "lead" variable) and heart rate ("hr" or "HR") as the second variable named in the CCF command (because it seems likely that heart rate may increase several seconds after an increase in talking). Note that because each observation corresponds to a 2-second time interval, we also need to be explicit in specifying whether the lead–lag relationship is expressed as a number of observations or as a number of seconds.

Initial examination of the two time series (amount of talk and heart rate) indicated that there was only a very small percentage of variance associated with a linear trend in either time series, so no trend removal was performed. If there had been a substantial trend component in either time series, the trend component should be removed from both time series before doing a CCF. Some analysts insist that it is necessary to prewhiten one or both time series, that is, to remove all forms of serial dependence such as lagged autocorrelation from one or both time series before looking at a lagged CCF. The issue of how removal of autocorrelation within each of the two time series changes the CCF will be addressed in a later section of this chapter. At this stage of the analysis, the raw time-series data (without any prewhitening) were used for the CCF.

The lagged CCF for amount of "FTALK" and "HR" (the raw time series, without removal of the trend or any other kind of serial dependence) is shown graphically in Figure 10.1. Note that the largest positive

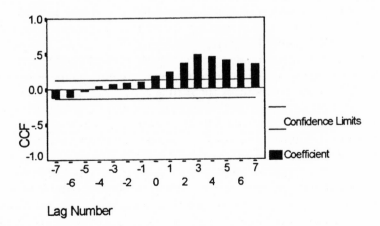

FIGURE 10.1. Lagged cross-correlation function between the amount of talk ("FTALK") and heart rate ("HR"); raw time-series data (Δt = 2 seconds).

cross-correlation occurred at a lag of three observations. For this lag of three observations, the correlation between "FTALK" and "HR" is almost +.50. This implies that an increase in talk tends to be followed by an increase in heart rate, with the maximum impact occurring about three observations (about 6 seconds) after the change in amount of talk. Based on only a correlational analysis, we cannot be certain that this indicates any causal influence; however, experimental studies in which the amount of talk is the manipulated independent variable have also found an increase in heart rate after the onset of talk. These better-controlled laboratory studies suggest that talking causes a small increase in heart rate.

However, the present data were not collected in a controlled experimental situation, and therefore it is possible that the lagged correlations are due to some spurious relationship. For instance, some third variable (like the behavior of the conversation partner) might first cause an increase in talk and then a few seconds later cause an increase in heart rate; the influence of some outside variable could produce the kind of CCF we see in this example. If we had time-series data on the outside variable that is suspected to be operating, such as partner activity, we could partial it out of the two time series using time-series regression analysis. We could then examine the relationship between the residuals of these time-series regressions to see if heart rate is still related to the amount of talk, even after partner activity is statistically controlled, to see if we could rule out partner activity as a source of spuriousness. However, I will not present empirical examples of such multivariate time series here.

From the lagged CCF it appears that the impact of talk on heart rate is spread out across time, such that it has some effect within one observation (2 seconds) and may continue to have some effect seven observations or more later (14 seconds or more).

Now we turn to the question of whether the cross-correlations between these variables might be partly (artifactually) due to serial dependence within one or both of these time series. For instance, we know that once people begin to talk, they tend to continue talking for a while. If this is the case, then it may be difficult to distinguish whether an increase in heart rate is due to the amount of talk at time t or to the amount of talk at times $t - 1$, $t - 2$, and so forth.

We can address this question by removing the serial dependence or autocorrelation from both time series before computing the lagged CCF; as noted earlier, removing serial dependence from a time series is often called prewhitening. Before we can remove serial dependence, we have to be able to describe it with a relatively simple equation or model. An autoregressive model is the simplest way of modeling serial dependence within one time series.

Let the amount of talk at time t be denoted X_t; the amount of talk one observation earlier is thus X_{t-1}; talk two observations earlier is X_{t-2}; and so forth. When we say that a time series of the amount of talk is "serially dependent" or autocorrelated, we mean that we can predict (to some extent) the current amount of talk from recent past amounts of talk. An equation that seems to fit well in many studies of serial dependence of vocal activity is the following second-order autoregressive equation:

$$X_t = \phi_0 + \phi_1 X_{t-1} + \phi_2 X_{t-2} + \epsilon_t$$

This equation is similar in many respects to an ordinary regression equation: ϕ_0 corresponds to the intercept term; ϕ_1 and ϕ_2 correspond to slopes or regression coefficients; and ϵ_t is a residual term. This is called a second-order autoregressive model because two lagged X terms are used to predict the present value of X_t. The equation simply says that, to some extent, X_t is predictable from the amounts of talk at immediately previous times, X_{t-1} and X_{t-2}. In most studies of vocal activity the ϕ_1 coefficient is fairly large and positive, which indicates that once a person is talking there is a tendency to continue talking.

The difference between autoregression and ordinary regression analysis has to do with assumptions about ϵ_t, the residuals. In ordinary regression analysis, we are required to assume that these residuals are uncorrelated (and for non-time-series data, often this assumption is satisfied). However, it is very often the case that time-series observations are serially correlated or nonindependent. Thus, if our time-series autoregression model shown in the foregoing equation does not account for all of the serial dependence in the time series, then these ϵ_t residuals will be correlated with each other. This violates the assumptions of independence that are required for construction of an appropriate error term for significance testing.

If we fit an autoregressive model to a time series (such as $X_t = \phi_0 + \phi_1 X_{t-1} + \phi_2 X_{t-2} + \epsilon_t$) and find that this model does not have independent residuals, then we need to modify the model to include whatever predictor variables are required to account for all the serial dependence in the time series. Usually this problem can be dealt with by including a few more lagged X terms as predictors, for instance, X_{t-3} and perhaps X_{t-4}. In some cases, more elaborate ARIMA models that include moving average terms or differencing might be required, but that is beyond the scope of this book; see McCleary and Hay (1980) for a more extensive treatment of ARIMA models that include moving average and differencing in addition to the autoregressive terms. Autoregressive models that include two lagged predictors are adequate to yield reasonably uncorrelated residuals in many behavioral time-series data (Warner, 1992b).

The SPSS TRENDS ARIMA program can be used to fit an autoregressive model of this form separately to each time series, and it creates several new variables based upon the analysis. The prewhitening strategy that was used to remove serial dependence from the heart rate and the talk time series was as follows:

1. Autoregression analysis was performed on the time series for talk (each observation is the proportion of time spent talking in each 2-second observation period). First-, second-, and third-order autoregressive models were fitted, and it was decided that the second-order autoregressive model (i.e., the equation that used X_{t-1} and X_{t-2} as predictors of X_t) was adequate. The residuals from this model were "white noise" or uncorrelated; this was confirmed by running a lagged autocorrelation function (ACF) on the residuals to make sure that there was no significant autocorrelation left. The coefficients for the second-order autoregressive model to predict talk at time t (denoted "$TALK_t$") from talk at previous times obtained from the SPSS TRENDS ARIMA program were as follows:

$$TALK_t = .61 + .58 \cdot TALK_{t-1} + .13 \cdot TALK_{t-2} + \epsilon_t$$

The residuals from this second-order autoregressive model, which were named "RESART," were saved as a variable in the SPSS worksheet. These residuals consist of the remaining component of the talk time series when virtually all serial dependence has been removed. Because these residuals are white noise, it is possible to do conventional significance tests for this regression. For the overall regression equation above, $\underline{R}^2 = .47$, $F(2,235) = 102.39$, $p = .001$. For the two regression coefficients, the t values were both statistically significant, $p < .05$.

2. Similarly, an autoregressive model was chosen for the heart rate time series; heart rate at time t was predicted from heart rate at times $t - 1$ and $t - 2$. The second order autoregressive model estimated by SPSS ARIMA was

$$HR_t = 62.83 + .70 \cdot HR_{t-1} - .4 \cdot HR_{t-2}$$

The residuals from this model were saved as the variable "RESAR2HR."

3. A lagged CCF was then run on the *residuals* from these autoregressive models (i.e., between "RESAR2T" and "RESAR2HR") to assess whether heart rate and variations in the amount of talk are still statistically related when serial dependence within each time series (in particular, the tendency for persons who are talking to continue talking) has been statistically controlled or partialed out.

The resulting CCF (between heart rate and the amount of talk, controlling for serial dependence within each time series) is presented in Figure 10.2. After controlling for the serial dependence within both time series, the lagged cross-correlations have become much smaller. However, the cross-correlation at a lag of three observations (6 seconds) is statistically significant (and the significance test is now valid because the residuals are statistically independent). This indicates that even after controlling for serial dependence there is still a tendency for an increase in the amount of talk to be followed about 6 seconds later by an increase in heart rate.

To get a better sense of the magnitude of change in heart rate, a simple regression analysis was performed using the residuals from the AR(2) model for talk to predict the residuals from the AR model for heart rate, at a lag of three observations. Recall that lagged ACFs were run on these residuals to confirm that these are now "white noise" or independent, and so there is no problem with the assumption of independence of residuals; therefore ordinary significance testing procedures may be applied. The ordinary linear regression procedure in SPSS was used to

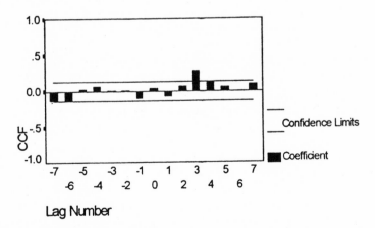

FIGURE 10.2. Lagged cross-correlation function between prewhitened time series for the amount of talk ("RESAR2T") and heart rate ("RESAR2HR") [prewhitened using AR(2) model].

obtained estimates of the intercept and slope, and to test the significance of this equation. The nonstandardized or raw score form of this regression model was as follows:

$$\text{RESAR2HR}_t = \phi_0 + \phi_1 \cdot \text{RESAR2T}_{t-3} + \epsilon_t$$

$$\text{RESAR2HR}_t = 0.00 + 5.00 \cdot \text{RESAR2T}_{t-3} + \epsilon_t$$

This equation indicates that an increase in the amount of talk from 0 to 1 (i.e., from not talking at all to talking all the time, during a 2-second time interval) is associated with an increase of about 5 beats per minute in heart rate, about three observations (6 seconds) later. This regression was statistically significant, $R^2 = .08$, $F(1,235) = 20.32$, $p < .001$. That the residuals from this model were not serially correlated was verified by looking at a lagged ACF for the residuals from this model to make certain that ordinary significance testing procedures could legitimately be used here.

None of the analyses presented so far address the question of whether either or both of these time-series variables tend to vary cyclically. This question may be addressed by doing a univariate spectral analysis on each time series (as described in earlier chapters). The results will not be presented in detail here. Heart rate did not show any statistically significant periodicity (a cycle on the order of 50 seconds long accounted for about 9% of the variance, but that was not statistically significant). The amount of talk did show statistically significant periodicity; together, cycles on the order of 4 and 8 minutes long accounted for about 23% of the variance in this time series. Thus, although this speaker's amount of vocal activity tended to vary in cycles longer than 3 minutes, there were no similar cycles in heart rate. Heart rate did tend to increase about 6 seconds after the onset of talking, but this apparent linkage between vocal activity and heart rate was not strong enough to make the heart rate also show cycles.

Because the univariate spectra showed no evidence of cycles in heart rate, it would not be useful to do a cross-spectral analysis for this pair of time series. The phase component of the cross-spectrum can be used, under certain conditions, to assess whether there is a lead–lag relationship between time series (see Gottman & Ringland, 1981, or a brief summary of their work in Chapter 9 of this book). However, in some research situations it is much easier to use the lagged CCF to address this question.

Another Empirical Example:
Detecting Coordinated Cycles in Talk

Another possibility is that there is no asymmetry or clear lead–lag relationship. A good example of this is variations in the amount of talk during a conversation. Speaker X tends to alternate between periods of mostly talking and periods of mostly listening. Partner Y tends to listen when X talks and talk when X listens, so partner Y's variations in amount of talk tend to be a mirror image of X's. When these variations in amount of talk are cyclic (and some research suggests that cycles about 3 minutes long in amount of talk occur in about half of all conversations), these cycles are in opposite phase between partners: that is, X is at the peak of the talk cycle when Y is at the trough. Each partner is a half cycle ahead (or, equivalently, a half cycle behind) the other, so it does not make sense to think in terms of an asymmetrical lead–lag relationship.

When we are dealing with a situation in which we are looking for coordinated cycles, we need to answer the following question in the analysis: What cycles are there in the activity of each person, X and Y? This question is addressed by looking at the two univariate spectra, as described in Chapter 6. This analysis will yield answers to such questions as these: Are there one or more major periodic components that are statistically significant? If yes, what is the frequency band (or cycle length) and amplitude for each major periodic component? The next major question is whether the cyclic activity within these frequency bands that correspond to major periodic components is statistically predictable between time series? This question is addressed by looking at the coherence spectrum (a component of the cross-spectrum). Squared coherence estimates the percentage of variance in time series X at a particular frequency that is predictable from variation in time series Y within that same frequency band. Note that high coherence within a particular frequency band can be interpreted as evidence of "synchronized cycles" *only* if there is enough power contained in that frequency band for each individual time series to warrant the interpretation that each time series actually has cycles at that frequency.

If high coherence occurs at some frequencies but the individual spectra do not suggest that there are any major periodic components or strong cycles within each time series, the data analyst should go back to lagged cross-correlation analysis to see if the relationship being detected is a lead–lag relationship that does not necessarily involve any cycling in either time series.

The last major question about coordinated cycles is the following: What is the phase relationship between these cycles? This question is addressed by looking at the phase spectrum (another component from the cross-spectrum). It only makes sense to look at the phase relationship between time series if there is high coherence between the series at that frequency; and it only makes sense to look at coherence between time series at a particular frequency if there is a substantial amount of power (explained variance) corresponding to that frequency band in the spectra of the two individual time series.

This empirical example illustrates coordinated cycles (in opposite phase). The vocal activity of two speakers (X = a female speaker; Y = a male speaker) in a conversation was monitored, and the proportion of time spent talking by each person in each 10-second observation period was recorded for 25 minutes. The complete data for these two time series ("FTALK" and "MTALK") are presented in Appendix A as "talkmf.sav." The data analyses and selected results for the relation between these two talk time series are presented in the subsequent paragraphs.

In order to assess whether each speaker tended to have cycles in her or his amount of talk, a univariate periodogram was performed for each of them. For each of these speakers, the frequency component that accounted for the largest percentage of variance in the time series was .02667 cycles per observation. Because cycle length is the inverse of frequency, this corresponds to cycles about 37.5 observations long; given that each observation was 10 seconds long, this corresponds to cycles about 375 seconds long, just over 6 minutes. Use of the Fisher test indicated that these cycles were statistically significant (α = .05). The tables and graphs for these univariate analyses are not presented here to save space, as they are the same kinds of tables as that presented in Chapter 6 (Table 6.1), but these would be included in a complete writeup.

Cross-spectral analysis was performed on this pair of time series to assess whether the vocal activity of these speakers was closely related, particularly within the frequency band that corresponded to these approximately 6-minute cycles. Graphs of the coherence and phase spectra are shown in Figures 10.3 and 10.4.

Coherence (which may be interpreted as, essentially, an $\underline{R}^2$ that estimates the proportion of variance in the Y (male) time series that is predictable from variance in the X (female) time series within each particular frequency band, or vice versa) was generally quite high across the low-frequency end of the spectrum. The frequency that is of most interest here is the one that accounted for most of the variance in each of the two individual time series; that is, the frequency of .02667 (in cycles per observation) or a cycle length of about 6 minutes. At this frequency, estimated coherence was .91, which suggests that about 91% of the variance in each speaker's vocal activity was predictable from the partner's vocal activity at this same frequency, at some time lag. A portion of the SPSS worksheet that shows the computed and saved variables that were created in the cross-spectral analysis is shown in Table 10.1.

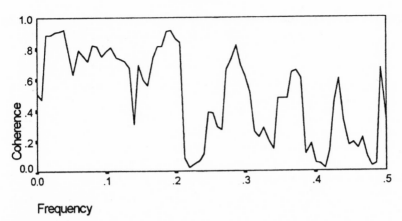

FIGURE 10.3. Coherence spectrum between female amount of talk, "FTALK," and male amount of talk, "MTALK" (Δt = 10 seconds). Note that the frequency (on the X axis) is given as cycles per observation. To convert this to frequency in cycles per second, divide the frequency shown on the X axis by Δt (10). To convert the cycles per observation into an estimate of the period (in number of observations, take the inverse (.5 cycles per observation corresponds to a cycle that is two observations, or 20 seconds, in length).

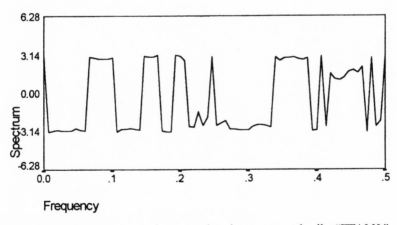

FIGURE 10.4. Phase spectrum between female amount of talk, "FTALK," and male amount of talk, "MTALK" (Δt = 10 seconds).

TABLE 10.1 Excerpt from the SPSS Worksheet Showing Some of the Results of the Univariate Spectral Analysis and Harmonic Analysis for the Male and Female Time-Series Data on the Amount of Talk during a Conversation

seconds	ftalk	mtalk	freq	fsin_l	fcos_l	fpdg_l	msin_l	mcos_l	mpdg_l
10	.93	.28	.00000	.00000	.58817	.00000	.00000	.33383	.00000
20	.73	.10	.00667	.08694	.07821	1.02558	-.07040	.00001	.37173
30	.75	.33	.01333	-.00301	-.03039	.06995	-.00125	.05929	.26376
40	.73	.23	.02000	.01871	.04936	.20902	-.01167	.00300	.01089
50	.75	.25	.02667	-.20651	.15062	4.89980	.19519	-.15043	4.55460
60	.75	.43	.03333	.09347	-.09154	1.28373	-.04429	.07887	.61362
70	.88	.50	.04000	-.01705	.01824	.04677	.03339	-.06215	.37329
80	.83	.35	.04667	.06345	.04418	.44835	-.06862	-.00727	.35711
90	1.00	.33	.05333	.03951	-.01871	.14334	-.05483	-.00844	.23080
100	.57	.43	.06000	-.03403	-.03226	.16492	.04485	.00430	.15225
110	.65	.63	.06667	.07928	-.02770	.52898	-.04023	.00771	.12581
120	.73	.25	.07333	.03785	.10237	.89341	-.03573	.06685	.43093
130	.45	.57	.08000	.03828	.00649	.11308	-.04647	-.4553	.31743
140	.03	.88	.08667	.00315	.04349	.14258	.03950	-.03237	.19561
150	.18	.70	.09333	-.08455	.03780	.64322	.06048	-.00517	.27632
160	.20	.88	.1000	-.03044	.06320	.36910	.01169	-.03767	.11666
170	.05	.80	.10667	.00890	.05327	.21879	-.01121	-.03830	.11945
180	.53	.45	.11333	-.00996	-.01760	.03066	.02377	.01722	.06460
190	.88	.00	.12000	.05286	.02034	.24058	-.07959	-.01104	.48424
200	.95	.25	.12667	-.02721	.00415	.05682	-.00958	.00004	.00689

Note. The variables "FTALK" and "MTALK" in the text and figures are shown as "ftalk" and "mtalk" in the column headings of this worksheet.

Because only the low-frequency end of the spectrum had large peaks that suggested major periodic components, results for only the first 20 periodic components are presented here. The estimated value for the coherence at a particular frequency can be approximated by looking at the graph of the coherence spectrum, but it can be estimated more precisely by scanning the coherence spectrum that was saved as a variable in the SPSS worksheet and reading the value there.

To estimate the time lag or phase relationship we now look at the phase spectrum, particularly at the phase estimate for the frequency band of .02667 (because this was the frequency band that accounted for a high proportion of variance in the vocal activity of each of the speakers). At this frequency, the phase relationship (reported by SPSS in radians) was −3.06666, which is almost exactly −π. If we divide by 2π to convert radians to proportions of a cycle, this means that the speakers' vocal activity peaks occurred almost exactly half a cycle apart. A phase relationship of +π radians or one-half cycle ahead is logically equivalent to a phase relationship of −π radians or one-half cycle behind; it is logically equivalent to say that speaker X is one-half cycle ahead of speaker Y or one-half cycle behind speaker Y when the cycles are repeating regularly. This suggests the 6-minute cycles for these speakers in this empirical example are synchronized in opposite phase; that is, when speaker X is most talkative, speaker Y is least talkative.

To illustrate this outcome graphically, the following analyses were performed.

1. The univariate spectral analysis was run for each speaker's vocal activity time series. The "save" subcommand in the SPSS SPECTRA command was used to save the coefficients for the sine and cosine components of the Fourier transform for each speaker.

2. The major periodic component that accounted for most of the variance in the vocal activity of both speakers corresponded to the frequency of .02667 (cycles per observation). This implies a cycle that is about 37.5 observations long. Because each observation in this time series corresponds to 10 seconds, the cycle length or period is approximately 375 seconds (just over 6 minutes). The Fourier coefficients for the sine and cosine components for each of the two time series (male and female talk) at this frequency were taken from the SPSS worksheet ("talkmf.sav") shown in Table 10.1; the names used for these were "fcos_1," "fsin_1," "mcos_1," and "msin_1."

3. An equation was set up to generate the sinusoidal component for each speaker that best fits the time series, using these sine/cosine coefficients. For the male speaker the predicted vocal activity cycle was estimated using the following equation, where t = the observation number:

$$MCYC6 = .19519 \sin(2\pi t \cdot .02667) - .15043 \cos(2\pi t \cdot .02667)$$

This new variable was named "MCYC6." When a cyclic component that has been identified through spectral analysis or periodogram analysis is plotted against the raw time series, it is helpful to add the mean of the time series (and also any trend component that has been fitted) so that the goodness of fit between the cyclic component and the raw time series can be seen more easily. The 6-minute cycles in amount of talk for the male speaker, with the mean level of talk added, is shown plotted against the raw time-series data for amount of talk for the male speaker in Figure 10.5.

Similarly, the 6-minute cyclic component in vocal activity is estimated for the female speaker using the Fourier coefficients from the SPSS worksheet:

$$FCYC6 = -.20651 \sin(2\pi t \cdot .02667) + .15062 \cos(2\pi t \cdot .02667)$$

This variable was named "FCYC6." This cyclic component, with the mean of the time series added to it, is shown plotted against the original time series in Figure 10.6.

Figures 10.5 and 10.6 show the 6-minute cycles (with the mean lev-

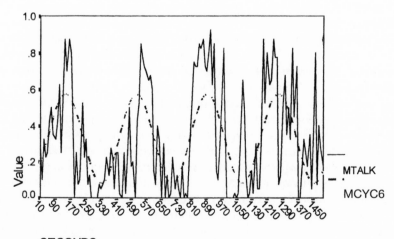

SECONDS

FIGURE 10.5. Amount of vocal activity for the male speaker, "MTALK," in 25-minute conversation ($\Delta t = 10$ seconds), with superimposed sinusoid ("MCYC6") corresponding to frequency of .02667 cycles per observation or 6-minute cycles.

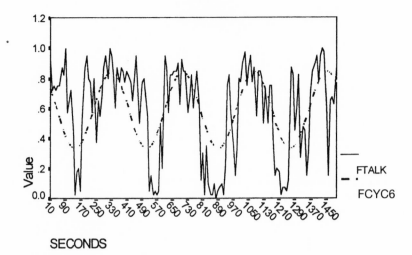

SECONDS

FIGURE 10.6. Amount of vocal activity for the female speaker, "FTALK," in 25-minute conversation (Δt = 10 seconds), with superimposed sinusoid ("FCYC6") corresponding to frequency of .02667 cycles per observation or 6-minute cycles.

el of talk added) that were fitted to the vocal activity of each speaker. For each speaker, the graph indicates that the amount of vocal activity tended to show relatively long cycles, on the order of 6 minutes long.

Figure 10.7 shows the phase relationship of these cycles between the male and female speakers. The 6-minute cyclic pattern in the amount of talk for the female speaker were plotted with the 6-minute cycles in the male speaker's vocal activity. (In this graph, the means were not added to the cyclic components.) Peaks in the female's amount of talk correspond to troughs in the male speaker's amount of talk; when she was talking most, he was talking least (and vice versa). This is consistent with the finding from the phase spectrum (see Figure 10.4), which showed that there was a phase relationship of either $+\pi$ radians or $-\pi$ radians between these two time series at this frequency (.02667 cycles per observation). In fractions of a cycle, this means that each speaker is either one-half cycle ahead (or, equivalently, one-half cycle behind) the other speaker. That is, the cycles are synchronized in opposite phase, with one person's maximum amount of talk corresponding to the other person's minimum amount of talk.

This information about the relation between the vocal activity of speakers X and Y can be converted into a lead–lag relationship in time units. Because the cycle is about 6 minutes long and each speaker is

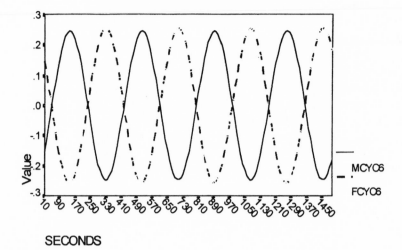

SECONDS

FIGURE 10.7. Fitted sinusoids corresponding to 6-minute cycles, in opposite phase, for male and female speakers, "MCYC6" and "FCYC6," respectively, in 25-minute conversation.

about one-half cycle "behind" the other speaker, this means that a peak in speaker X's vocal activity tends to be followed by a peak in speaker Y's activity about ½ cycle (or about 3 minutes, in this case) later.

Of course, other phase relationships between time series are possible. If we had studied two time series that tend to covary together, such as systolic and diastolic blood pressure time series, we would be more likely to find a phase near 0, which would indicate that the peaks in systolic blood pressure correspond to the peaks in diastolic blood pressure.

Chapter Summary

Many different outcomes of cross-spectral analysis are possible; the examples here illustrate two of the more interesting possible outcomes. For the analysis of talk and heart rate, a 6-second lagged relationship was found such that heart rate tended to increase 6 seconds after speaking. For the analysis of talk patterns in a male–female conversational dyad, it was found that both speakers showed approximately 6-minute-long cycles in amount of talk and that these cycles were synchronized in opposite phase (with one speaker talking most when the other speaker was

talking least). Other outcomes are possible, and some outcomes are suffi-
ciently complex and messy that they are difficult to interpret and report.
As in earlier chapters, I suggest using the simplest possible analysis;
sometimes when cross-spectral analysis results are difficult to interpret, it
may be better to go back to simpler methods, such as the lagged CCF, to
get a more understandable description of the data.

CHAPTER 11

Pitfalls for the Unwary: Examples of Common Sources of Artifact

Introduction

It is important to understand that, although a peak in a univariate periodogram may indicate the presence of a sinusoidal cycle in the time series, there are many ways in which spurious or misleading peaks can arise in a periodogram or power spectrum. One way to assess whether a peak in a periodogram is being misinterpreted is to plot the predicted sinusoidal component against the raw time-series data and examine this graph carefully to see what features of the time series are being captured by the analysis, as described in earlier chapters. However, it is also useful to be aware of the types of spectra that are likely to be obtained when the time-series data contain some common patterns such as a trend, spikes, or a "boxcar" or rectangular waveform. It is also important to understand "leakage" (see Chapter 5), which occurs when the cyclic components that are fitted to the time-series data in the spectral analysis do not have exactly the same cycle lengths as those in the time-series data.

Examples of Artifacts

Subsequent sections of this chapter present simulated time-series data to illustrate what the raw time series look like in each of these situations and to show what the corresponding univariate spectra look like. Keeping these examples in mind will help data analysts to avoid mistakes in interpretation of univariate spectra. A simulated "vocal activity" time series with an N of 150 observations was generated using the random number function in SPSS; a normally distributed random variate was created, and it was scaled so that its range of values would be similar to the data presented in earlier chapters with a minimum of 0 and a maximum of 1.00. This raw time series is given in Appendix A as a file named "simtalk.sav." (Actually, the SPSS random number function does not do a very good job of generating a truly random sequence of numbers; many "random number generators" use algorithms that contain some slight serial dependence in the scores that are generated, and so the "random" data will not look exactly like white noise in the analyses that follow.)

The first analysis that is presented in this chapter shows the periodogram that is obtained for this white noise time series. Each subsequent section of this chapter shows how the appearance of the raw time series changes when one of several types of pattern are added to the time series; then the periodogram is presented, to show that patterns such as a trend or spikes can generate artifactual peaks in a periodogram, that is, peaks that do not correspond to a true, regularly repeated sinusoidal cycle in the original time series.

A graph of this white noise random time series is shown in Figure 11.1, and the periodogram for this time series is shown in Figure 11.2. The periodogram for this time series of "random numbers" does not have any distinctive large peaks in it. The ideal spectrum for a true white noise process would be a flat line indicating that power is uniformly distributed across all frequencies. In any empirical example, some sampling error occurs, and so instead of the analyst seeing a periodogram plot that

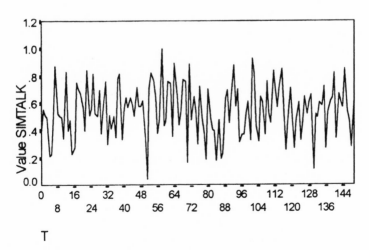

FIGURE 11.1. Simulated talk time-series data (from random number generator function in SPSS); T = time (in arbitrary units).

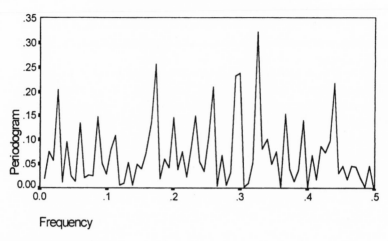

Frequency

FIGURE 11.2. Periodogram of SIMTALK, simulated talk time-series data (raw data shown in Figure 11.1).

is a flat line, the values of the observed periodogram tend to vary randomly around a flat line. The key thing to notice is that, when the proportion of variance accounted for by each of the $N/2$ individual frequency components in this periodogram are tested for significance (using the Fisher test described in Chapter 5), none of the individual values are statistically significant. Thus, there is no indication from the periodogram shown in Figure 11.2 that any one frequency component accounts for a much larger proportion of the variance in the time series than do most other frequency components.

Effect of a Linear Trend on a Univariate Periodogram or Spectrum

A second simulated time series (SIMTRND) was created by adding a linear trend component to the simulated talk time-series data from Figure 11.1 and rescaling the time series so that its range was approximately 0 to 1.0. The raw time-series data with this linear trend component is shown in Figure 11.3; and the periodogram that is obtained from this time series (when the trend is not removed) is shown in Figure 11.4.

The strong linear trend component in the Figure 11.3 time series produces a large peak at the low-frequency end of the periodogram shown in Figure 11.4. If other low-frequency cyclic components had been present, it would be difficult to distinguish them from the peak in

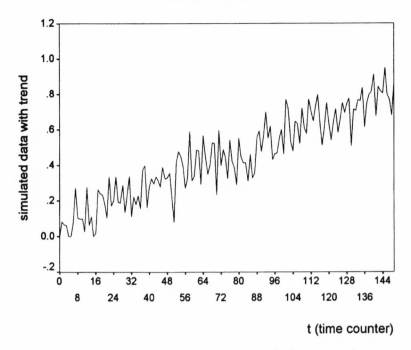

t (time counter)

FIGURE 11.3. Simulated talk time-series data with linear trend component added.

the periodogram that was created by this trend component. The point of this example is that, before a peak at the low-frequency end of the periodogram may be confidently interpreted as evidence for a relatively long cycle, we must check to see that it is not actually due to a linear or curvilinear trend component in the time series. In fact, when there is a linear or curvilinear trend in a time series, we really do not know whether this might be a piece of a longer cycle (i.e., a cycle with a period longer than the data record). A much longer time-series data record is needed to assess whether an observed trend might be due to cycles that are longer than the N in the original time-series data.

Effect of Spikes in the Time Series
on the Univariate Periodogram

Many familiar statistics (such as the mean of a sample or the Pearson correlation) are not robust to outliers; one or two extreme outliers can greatly alter the estimate of a mean or a correlation. In a similar way, pe-

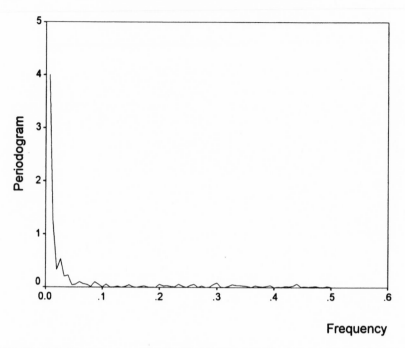

FIGURE 11.4. Periodogram of SIMTRND, simulated talk time-series data with linear trend component added (raw data shown in Figure 11.3).

riodogram analysis is not robust to the effects of outliers; a few extreme outliers in the time series can have a major impact on the shape of the periodogram and can produce patterns in a periodogram that are difficult to interpret even if the number of observations in the time series is large.

To illustrate this problem, two large outliers were added to the simulated vocal activity data (from Figure 11.1). A value of 5 was added to observations at times 16 and 31. The data with these spikes added are shown in Figure 11.5, and the periodogram that is obtained from this time series that includes two extreme outliers is shown in Figure 11.6.

The periodogram that is obtained when the time-series data include these two extreme outliers has a series of peaks. Although a periodogram that looks like this might appear to suggest a set of superimposed cycles of various lengths, in fact this pattern is due to the presence of the two extreme outliers. The point of this example is that extreme outliers (whether they represent real values or measurement errors) can have a disproportionate impact on the obtained periodogram and may make the periodogram extremely hard to interpret. Before spectral analysis is per-

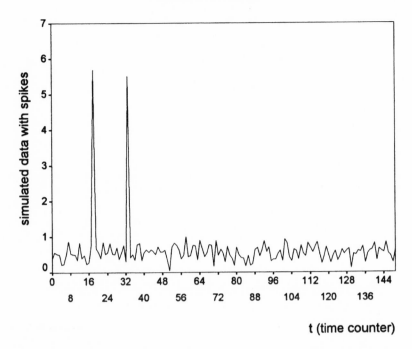

FIGURE 11.5. Simulated talk time-series data with two spikes added (extreme outliers, spaced 15 observations apart).

formed, extreme outliers need to be removed or modified (perhaps by log transformation of the raw time series or by trimming extreme values; refer to Chapter 2 for a discussion of outliers in time series).

Effect of a "Boxcar" or Rectangular Waveform on the Periodogram

It is common in time-series studies to include baseline periods before and sometimes after a treatment or observation period. If time-series data are collected during these baseline periods, the time series may end up having a kind of "boxcar" or rectangular waveform as one of its major components. Figure 11.7 shows the simulated vocal activity time series (from Figure 11.1), but with the first 15 and last 15 observations set to zero. Such a time series might be obtained in an actual study in which the subject is instructed to be silent for 15 observations, then to talk for 120 observations, then to be silent again for 15 observations. Similarly, if blood

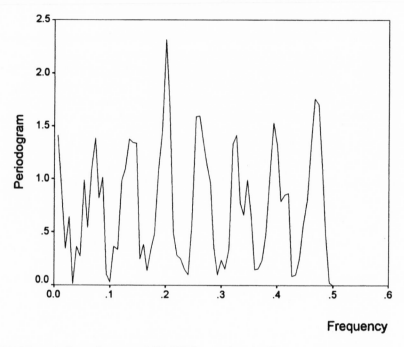

FIGURE 11.6. Periodogram of SIMSPK, simulated talk time-series data with two spikes added (raw data shown in Figure 11.5).

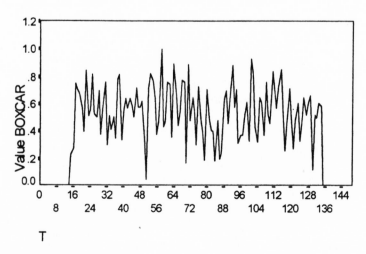

FIGURE 11.7. Simulated talk time-series data padded with zeroes to include "box-car" or rectangular waveform.

pressure is measured before, during, and after a stressor, the levels of blood pressure may shift drastically from baseline to treatment to return to baseline, which would also produce a kind of rectangular component in the time series. If this rectangular component is included in the analysis (i.e., if a periodogram is performed on all $N = 150$ observations including the baseline periods at the beginning and the end), the results will look similar to the periodogram shown in Figure 11.8.

The periodogram shown in Figure 11.8 has a peak at the low-frequency end. In effect, the "rectangular" waveform in the original time series was similar enough to a very long sinusoid that it gave rise to a peak that looks something like an indication of a long cycle in the data. This pattern could obscure real cyclic components in the time series.

The point of this example is that baseline periods (with very different mean levels than the treatment period) should be analyzed separately from the treatment period if the goal of the analysis is to detect any relatively long cycles in the time series. Alternatively, if the data analyst wants to do a spectral analysis that combines the baseline and treatment periods, the means of each time-series segment can be sub-

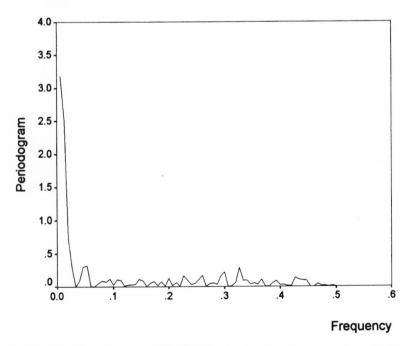

FIGURE 11.8. Periodogram of BOXCAR, simulated talk time-series with "boxcar" waveform (raw data shown in Figure 11.7).

tracted from each segment to remove any boxcar pattern from the time series.

Leakage That Occurs When *N* Is Not an Integer Multiple of the Cycle Length

The problem of leakage was discussed more extensively in Chapter 5, where examples showed that a 7-day cycle in mood is not accurately extracted by a periodogram analysis if the time series length is not an integer multiple of 7. Another example is provided here. Figure 11.9 shows the simulated talk time series with a 15-observation-long cycle superimposed on it (as in the earlier simulations in this chapter, the data were rescaled to have a range from about 0 to 1.0). Visual examination of this raw time series indicates that there are peaks spaced about 15 observations apart. The length of the time series, $N = 150$, is an integer multiple of this cycle length.

Two periodogram analyses were performed using the data shown in Figure 11.9. The first periodogram, shown in Figure 11.10, used all $N = 150$ observations. The Fourier frequencies included in the periodogram analysis were thus 1/150, 2/150, . . . , 75/150, and the corresponding cycle lengths thus included a cycle 15 observations long. This periodogram

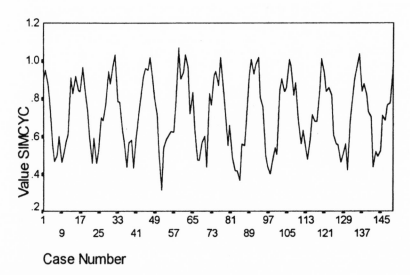

Case Number

FIGURE 11.9. Simulated talk time-series data with superimposed sinusoidal cycle 15 observations long.

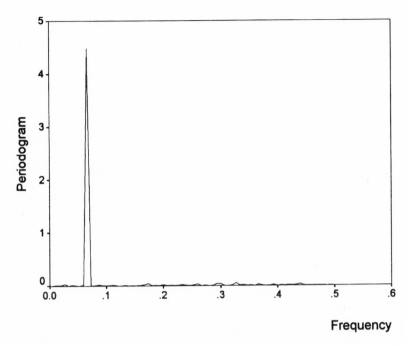

Frequency

FIGURE 11.10. Periodogram of SIMCYC, simulated talk time-series data with cycles 15 observations long (raw data shown in Figure 11.9). This periodogram illustrates the interpretable results obtained when the time-series length is an integer multiple of the cycle length. The peak in this spectrum corresponds to a frequency of .06667 cycles per observation, or an estimated cycle length of 1/.06667 = 15 observations.

has one very large peak that corresponds to a frequency of .06667 cycles per observation, or a cycle length of 15 observations. No leakage occurs here; the periodogram accurately detected the major periodic component and provided a good estimate of the cycle length.

A second periodogram analysis was done on a subset of the time-series data from Figure 11.9 to show how leakage can occur when the number of observations in the time series is not an integer multiple of the cycle length (and particularly when the N is not large, relative to the length of the cycles in the time series). For this periodogram analysis, only the first 68 observations of the time series shown in Figure 11.9 were included. Because this N (68) is not an integer multiple of the cycle length (15), leakage is a problem. The Fourier frequencies for this periodogram do not include a component that corresponds to a cycle 15 observations long. Because of this omission, the neighboring frequencies account for a relatively large share of the variance in the time series. There is still one fairly distinct peak in the periodogram, but the peak in the Figure 11.11 periodogram (with leakage) is much broader than the peak in the Figure 11.10 periodogram (without leakage).

If the data analyst tries to estimate the cycle length from a periodogram that suffers from leakage, the estimate of the cycle length will be inaccurate. The point of this example is that, once there is some basis to judge the cycle length—either from a priori knowledge, such as past research findings of 7-day cycles in mood, or from exploratory analyses and careful examination of the data—it is important to make the number of observations in the time series an integer multiple of the length of the cycle that we are trying to detect. (Harmonic analysis, introduced in Chapter 4, can also be applied to the time series, to assess the goodness of fit of various candidate cycle lengths. This can be done either as a preliminary to spectral analysis, to assist in choosing the Fourier frequencies in a manner that will avoid leakage, or as a follow-up to spectral analysis, to illustrate just how well the cycle length that was identified as important in the spectral analysis actually fits the raw time-series data—or both.) Furthermore, it is desirable to have as many repetitions of the cycle in the data record as possible. As a general rule,

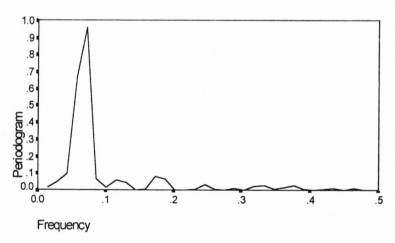

Frequency

FIGURE 11.11. Periodogram of SIMCYC, simulated talk data with cycles 15 observations long (using only the first 68 observations of the raw time-series data shown in Figure 11.9). This illustrates "leakage," a problem that occurs when the time-series length is not an integer multiple of the cycle length. The broad peak in this periodogram suggests estimated cycle lengths ranging from about 10 to 20 observations, which does not match the 15-observation cycles actually in the raw time-series data.

I would suggest the number of observations in the time series should be at least 5 (or, better yet, 10) times the length of the cycles that you are trying to detect.

Chapter Summary

In nearly every case, the best way to assess whether these possible sources of artifact are present is to go back and make a careful assessment of the original time-series data, to see what patterns in the data are being captured by the periodogram analysis. It is important to check that peaks in a periodogram or spectrum are not due to one of these other patterns (e.g., trend, spikes, boxcar waveforms, or leakage from other frequencies) before interpreting peaks as evidence of regular cycles.

An exception is the problem of aliasing, which cannot be detected by post hoc examination of the data. Aliasing (a problem that was first mentioned in Chapter 2) occurs when the behavior contains very high frequencies (very short cycles) but the sampling intervals are spaced far enough apart that they do not detect these rapid cycles. Instead, spurious "long" cycles appear in the data record because of the way the sampling frequency interacts with these underlying rapid oscillations. It is only possible to assess whether aliasing is a problem by sampling more rapidly to see if shorter cycles appear at this higher sampling rate.

To summarize: A time-series data analyst needs to evaluate whether any of these possible sources of artifact may be present before concluding that a peak in a periodogram or spectrum is a clear indication of cyclicity. Most of the problems noted in this chapter can be avoided by careful planning in the design stage and preliminary data screening (see Chapter 2). The sampling frequency and time-series length should be chosen in a way that makes it possible to see cycles if they really do exist, and in a way that avoids artifactual distortions of apparent cyclicity such as aliasing and leakage. If high-frequency (very rapid) cycles exist in the behavior that is being studied, then the researcher either must remove these high frequencies by filtering the data signal before taking samples (many polygraph modules or data acquisition programs provide such filtering) or must sample at a fast enough rate to capture these high-frequency cycles. The N of the time series should be an integer multiple of the cycle length that the researcher is trying to detect. Past research may suggest what cycle lengths are likely to occur, or careful examination of the raw time-series data and preliminary analysis using harmonic analyis may help the researcher determine what cycle lengths the design should be optimized to detect.

In addition, the graph of the time series should be examined to see if there are patterns such as linear trends, boxcar-shaped waveforms, or spikes (extreme outliers). Any of these patterns can produce artifactual peaks in a periodogram or power spectrum, and so each of these patterns must be removed from the time-series data before a periodogram or spectrum is estimated. If patterns such as trends or spikes are not removed from the time series, then the periodogram or power spectrum may be difficult to interpret.

CHAPTER 12

Theoretical Issues

Introduction

The statistical methods outlined in the first 11 chapters of this book provide ways of describing patterns in naturally occurring time-series data. These methods can be used to describe the proportion of variance in a time series that is attributable to trends and cycles, and to describe the nature of any cycles, such as the cycle length or period and the amplitude. However, researchers often want to go beyond mere description of a pattern to make various kinds of inferences—for instance, inferences about the factors that generate patterns in time series or factors that cause changes in the pattern of a time series. It is not generally possible to make such causal inferences unless an experimental time-series study has been conducted.

A complete description of the design and analysis of time-series experiments is beyond the scope of this book (see Gottman, 1981, and McCleary & Hay, 1980, for introductions). However, general issues about making causal inferences and some useful sources for the design and analysis of time-series experiments will be discussed briefly in this final chapter.

Questions a Researcher Might Seek to Answer

In univariate time-series research, a researcher often wants to make inferences about factors that may cause or modulate the periods and amplitudes of cycles. Analogous research questions arise in biological rhythms research, where researchers often want to demonstrate that observed cycles in physiology are at least partly due to endogenous factors (i.e., processes within the organism, such as "biological clocks" or more complex forms of temporal organization arising from complex organization of neural systems). The usual research strategy involves observation of the physiology of an organism that is in an unchanging environment, or "free-running" conditions. If a circadian rhythm in body temperature or motor activity persists in the absence of a 24-hour light–dark cycle or other time cues, then it is reasonable to infer that it may be at least partly generated by endogenous processes (regulatory mechanisms within the organism). Behavioral or social science researchers might want to do similar controlled-environment studies to see whether cycles in mood or behaviors persist when the environment is kept constant. Research handbooks that describe the design and analysis of free-running biological rhythms provide useful suggestions for the design and analysis of this type of research (e.g., see Aschoff, 1981).

Researchers are also often interested in the ways that cycles are modulated by environmental events; for instance, introducing a change in the light–dark periodicity in the environment can result in a change in the phase and/or period of the biological rhythm. Inferences about causes of cycles cannot be made from purely observational studies of time series. As in any other research situation where the goal is to make a causal inference, the researcher needs to conduct studies that include experimental manipulations and controls over extraneous variables. If a researcher wants to know whether a particular type of transient disturbance or environmental event changes cycles in a specific way, carefully controlled experiments are needed in which the environmental "disturbance" is introduced while other extraneous variables are controlled.

Another type of inference researchers are often interested in involves the possible causal impact of one time series (X) on a second time series (Y). For example, a researcher may want to know whether changes in an infant's behavior (represented by an X time series) cause subsequent changes or adjustments in the parent's behavior (represented by a Y time series). It has been suggested by some time-series analysts that a time series regression analysis in which the researcher predicts values of Y_t from lagged values of X (such as X_{t-1}, X_{t-2}, ...), while controlling for or partialing out past values of Y (Y_{t-1}, Y_{t-2}, ...), provides her or him with a sufficient basis for making a causal inference. If lagged values of X predict Y even when lagged values of Y are partialled out, then some possible sources of spurious correlation between the X and Y time series have been statistically ruled out. However, as discussed in Chapter 8, this time-series regression analysis of purely observational data does not provide a strong basis for a causal inference. A better case for causal inference could be made by manipulation of the X time series and observation of subsequent changes in the Y time series, while other extraneous variables are controlled. (Of course, such interventions are not always feasible, but experimental time-series studies are often desirable in research situations where they are ethically acceptable and also practical.)

The experimental introduction of an intervention to assess its impact on an ongoing time series can be done in several different ways. A possible approach is (1) to collect baseline time-series data in the absence of an intervention; (2) then to introduce either a "one-shot" intervention that consists of a brief event or a longer-lasting intervention that consists of a continuing treatment; and (3) then to assess the nature of the impact of this intervention on the behavior of the time series. An excellent introduction to the design and analysis of interrupted time-series experiments is provided by McCleary and Hay (1980). The researcher may look for various types of impact: immediate or delayed; temporary or permanent. Most researchers look at changes in the level or mean of the time series (comparing the pre- and postintervention periods), but it is also possible to look for changes in other features of the time series, such as its variance, trend, or cycles.

Another somewhat more complex method for assessing the effect of interventions was proposed by Gregson (1983). In this approach, the X time series is "created" or generated entirely by the experimenter (and thus it may be a white noise or random series of events) and the Y time series consists of observations of responses that are made by a subject. For example, the X time series might consist of a series of stimuli (such as noises of varying loudness), whereas the Y time series would consist of magnitude estimation responses. The simplest situation would be to create an independent X time series that is white noise, that is, uncorrelated or unpatterned. In this situation, it is relatively easy to assess the impact of a change in the X time series on the Y time series, to detect lead–lag relationships, to assess whether the impact is long lasting, and so forth. This research paradigm (which makes sense for studies in psychophysics) might possibly be adaptable to research on social behavior, although the occurrence of "random" social behavior would probably seem very strange to the research subject and it might be difficult to create an intervention time series of social behaviors that did not seem unnatural.

It is not always feasible to manipulate one of the time series, but when an experimental design is possible, this will generally provide a stronger basis for causal inference than the statistical controls that are employed in bivariate time-series regression analyses.

Is the Underlying Process Stochastic or Deterministic?

Another type of inference researchers often want to make concerns the nature of the underlying process that generates the observed behavior time series. Can the pattern in the time series be represented by a mathematical or statistical model? If so, should the model presume a deterministic or stochastic process? What type of equations are needed, and what parameters are involved? These questions are discussed by Buder (1996).

At this point, efforts at formal model identification for behavioral time-series data are still at a very early stage (e.g., see Buder, 1991, and Cook et al., 1995). Unfortunately, most time-series data in the behavioral sciences have substantial errors of measurement and relatively small numbers of observations; this makes accurate model identification difficult. Furthermore, comparison of simple measures of goodness of fit (such as the magnitude of sum of squares residuals, or R^2) does not necessarily lead to identification of the "best" model. A good model should do more than simply fit the data; it should also generate testable predictions, and the parameters of the model should be interpretable in ways that are theory relevant. (For excellent examples of this, see model development work by J. M. Gottman and his colleagues, reported in Cook et al., 1995.) Given a mathematical or statistical model for a time-series process, it is sometimes possible to generate empirically testable predictions from that model. (Winfree, 1983, has done this with exceptionally useful results in modeling the cardiac cycle; his model correctly predicted that a transient disturbance at a critical point in the cardiac cycle would induce fibrillation, a pathological breakdown in normal heart rhythm, and this theoretical prediction has been experimentally verified.)

However, in many social and behavioral research situations, we do not yet have enough knowledge or theory to propose reasonable mathematical models. Unfortunately, most behavioral science researchers are still quite far from being able to propose specific mathematical models that make testable predictions that would allow discrimination among models, although some extremely promising pioneering work has already been done (Buder, 1991; Cook et al., 1995). It is important to remember, however, that mere goodness of fit is not a sufficient criterion for choosing a model (this has been pointed out in classic sources on time-series modeling, such as Box & Jenkins, 1970, but it also holds true for other kinds of mathematical modeling).

Reasonably readable introductions to some mathematical models that have proved to be useful in describing some biological processes over time, and that might also (with suitable modifications) be useful models for time series on mood and behavior, are presented in Murray (1993) and in Glass and Mackey (1988). While relatively simple linear systems models may work reasonably well to describe some behavioral systems, many researchers are now turning to somewhat more complex nonlinear systems models for better description (see Glass & Mackey, 1988, and Buder, 1991, for good introductions to nonlinear systems).

Why Cycles Might Arise in Behavior and Physiology

Many physiological processes are now known to show cycles (24-hour circadian cycles, but also cycles that are much longer and much shorter; see Webb, 1982, and Moore-Ede, Sulzman, & Fuller, 1982). Many of these cycles persist even when environmental variations such as the 24-hour light–dark cycle are experimentally controlled or held constant; this has led to the speculation that there is some sort of internal "biological clock" that generates these cycles (or perhaps something more like a "clockshop," given how complex the rhythmic patterns are now suspected to be; see Winfree, 1975). There is no consensus about the physiological basis of endogenous timing; some researchers have focused on the existence of a "clock" gene or on subcellular rhythmic processes; others note the importance of complex organizational properties of the central nervous system, and particularly the importance of the suprachiasmatic nucleus. Whatever endogenous regulatory processes may generate these cycles, researchers do agree that these cycles can be modulated and/or disrupted by environmental events, so they are not exclusively biologically driven.

The existence of biological rhythms may not be a mere curiosity, but rather a fundamental organizing principle of living systems (Iberall & McCulloch, 1969). Theorists in biology, such as Goodwin (1970), have pointed out that cycles are a dynamic form of equilibrium that is more suited to living systems than is the static sort of equilibrium observed in some simple physical systems.

A simple example of a system with a static equilibrium is a thermostat that is set to a constant temperature level; when the room temperature goes above this threshold, the thermostat turns the furnace off; when the room temperature goes below this threshold, the thermostat turns the furnace on. The effect is that the room temperature is maintained at a nearly constant level.

An example of a dynamic equilibrium is body temperature. Most organisms, including human beings, do not have a constant body temperature; instead, the body temperature shows a circadian (about 24-hour) cycle, with higher body temperatures during the more active daytime hours and lower body temperatures at night. When momentary events (such as engaging in exercise) change the body temperature, the body's homeostatic mechanisms operate in a way that tends to make temperature return toward this baseline 24-hour cycle.

Dynamic equilibrium (cycles) has some advantages over static equilibrium (Goodwin, 1970). First, the environment tends to vary periodically (most noticeably, the 24-hour light–dark cycle, and the yearly seasonal variations in temperature). Circadian and annual rhythms in physiology allow the organism to adapt to, and even to anticipate, these environmental variations. Second, because so many internal physiological processes need to be coordinated with each other, synchronization of internal physiological rhythms may be a key mechanism involved in maintaining order in the internal physiological milieu; for instance, Goodwin (1970) noted that the timing of cell wall division needs to be coordinated with the timing of chromosome division, and even at subcellular levels many physiological processes are now known to be cyclically or rhythmically organized. Finally, coordination of physiological rhythms between organisms may be an important link among components of social systems. Many animals coordinate their estrous cycles and birth of young; there is some evidence suggesting that under certain conditions women coordinate their menstrual cycles (McClintock, 1971) and that women and men in relationships may coordinate certain hormonal cycles (Kruse & Gottman, 1982). Field (1985) speculates that the entrainment of biological and behavioral rhythms between infants and their caretakers may be an important component of attachment and development. Hofer (1984) has further argued that social relationships serve a crucial function in modulating and regulating biological rhythms and that loss of this regulation is one of the factors that make bereavement or isolation so devastating to physical health. Healy and Williams (1988) are among several theorists who now believe that some forms of depression may be—at least in part—due to the desynchronization or disruption of normal physiological, emotional, and behavioral rhythms.

Thus, there is a growing amount of both theory and data supporting the idea that rhythmic or cyclic organization is important for the "healthy" functioning of all sorts of living systems—organisms, but also social systems. For more discussion of these issues, see Warner (1988).

Are Some Forms of Rhythmic Organization More Likely to Be "Optimal" Than Others?

An implicit theme in some of the early biological rhythms research seemed to be that "rhythm is good." Factors that disrupt normal cycles, such as the effects of jet lag or swing-shift work on human circadian rhythms, were noted as problems that could have a detrimental effect on both physical and mental health (Moore-Ede et al., 1982). However, not all rhythms or cycles are "healthy". Certain types of rhythms, such as a periodic apnea or cessation of breathing called Cheyne–Stokes respiration, are in fact pathological. Some theorists have gone to a different extreme, arguing that "chaotic" patterns in physiological processes such as the cardiac cycle are in fact more conducive to good health than are regular cycles (Pool, 1989).

At this early stage in research, it would be a gross oversimplification to assume that regular cycles are always an indication of better functioning. In fact, the degree of predictability or cyclicity that is optimal may vary depending upon what behaviors or physiological processes are involved or on the nature of the social relationship being studied (Cappella, 1988; Warner, 1992b). A great deal of empirical work will be needed to assess what kinds of cyclic patterns and what strength of coupling between time series are optimal in different research situations. However, there are already several studies that suggest that asking such questions is fruitful.

For instance, the strength of the coupling between a person's respiration and heart rate can be assessed by looking at respiratory sinus arrhythmia—variations in heart rate that reflect the effects of respiration on the heart. This coupling is now often called "vagal tone" because the vagus branch of the parasympathetic nervous system mediates this coupling between respiration and the cardiac cycle. Porges and Coles (1982) have found that there are individual differences in the strength of this coupling. There is now considerable interest in vagal tone (which can be assessed using cross-spectral analysis or more task-specific analytic methods suggested by Porges and Coles) as an important individual physiological difference that has implications for temperament (Fox, 1989).

Some research on "cyclicity" (or within-subject predictability of behavior over time) in social interaction has examined how the strength of cyclicity or predictability relates to social outcomes, such as person perception, evaluation of the social interaction, and liking for the partner (Warner, 1992c). Results have been somewhat mixed: overall, it appears that stranger dyads may evaluate their social interactions somewhat more positively when the interaction is more cyclic, more predictable, and more coordinated, whereas couples who know each other well, such as married couples, may evaluate their interactions more positively when they are less cyclic, less predictable, and less coordinated (Warner, 1996). This is consistent with suggestions by Berger and Bradac (1982) that predictability serves to reduce uncertainty. Uncertainty reduction may be welcome in a relatively anxiety-provoking situation such as interacting with a stranger. On the other hand, some unpredictability may provide welcome novelty in interactions with persons who know each other well. At this point, it is fair to say the following:

1. The degree of predictability and strength of coordination (which can be assessed by using periodogram analysis and cross-spectral analysis and other methods described in this book) may provide important information about how "well" the system is functioning (although it is not always the case that more predictable and more strongly coupled patterns are "better").

2. The degree of predictability that is optimal probably varies substantially depending upon the variables that are being studied (in physiology/behavior research) and the nature of the task and the social relationship between participants (in social interaction research).

The Relation between Frequency-Domain and Time-Domain Analysis of Time-Series Data

There are two related methods for analysis of time-series data. The first analytic approach, and the one that has been the primary focus of this book, includes frequency-domain methods such as harmonic analysis, periodogram analysis, and spectral analysis. These frequency-domain analyses involve decomposing the variance of the time series into variance that is accounted for by a set of sinusoidal cyclic components. The second analytic approach is the set of time-domain methods formally called Box–Jenkins ARIMA modeling (Box & Jenkins, 1970); here AR stands for "autoregressive," I stands for "integrated," and MA stands for "moving average." Autoregressive processes, which provide a relatively simple and very useful method for describing a pattern in time-series data and also a means of "whitening" or removing that pattern from time-series data, were discussed in Chapter 8. Integrated and moving average models, which may also be useful in many situations, were not included here; see McCleary and Hay (1980) or Box and Jenkins (1970) for complete treatments of these topics.

Time-domain methods, such as autoregressive models for time series (e.g., equations that predict X_t from X_{t-1}, X_{t-2}, and perhaps other lagged values of X), are an alternative way of representing information about patterning in time series. Mathematically, frequency-domain and time-domain representations contain equivalent information about the time series (this is demonstrated formally by Box & Jenkins, 1970).

For data analysts, this mathematical equivalence between the frequency-domain methods covered in this book and the time-domain methods that are also used extensively by researchers has some interesting consequences. For example, for certain specific values of the autoregressive coefficients ϕ_1 and ϕ_2, the time series that is generated by a second-order autoregressive process ($X_t = \phi_0 + \phi_1 X_{t-1} + \phi_2 X_{t-2}$) tends to show relatively long although not very regular cycles (see Box & Jenkins, 1970, p. 59, for details). Thus, a time series that is well explained by certain kinds of second-order autoregressive models, and therefore has a large $\underline{R}^2$ for the fit of these AR(2) models to the time-series data, will also tend to have a rather large and broad peak at the low-frequency end of the spectrum—or, in other words, a relatively high percentage of the variance in the time series accounted for by long cycles. It is important to note that a large autoregressive $\underline{R}^2$ does not necessarily imply periodicity (a peak near the low-frequency end of the spectrum); whether strong autocorrelation suggests cyclicity or other types of pattern in the time series depends upon the actual values of the ϕ_1 and ϕ_2 coefficients.

Note that a researcher's preference for use of frequency-domain versus time-domain data analysis is typically influenced by the researcher's theoretical background. Researchers who are primarily interested in the detection and description of cycles tend to use spectral analysis and related frequency-domain techniques, as described in this book. Researchers who are primarily interested in removing a pattern from time series by prewhitening, in forecasting future values of time series, and in testing hypotheses in interrupted time-series experiments tend to choose time-series regression and other time-domain analytic techniques related to the Box–Jenkins ARIMA modeling approach.

However, it is important to understand that these techniques actually detect many of the same kinds of pattern in time-series data, and to understand that these methods cannot be used to distinguish between "deterministic" versus "stochastic" cycles. If spectral analysis yields a large peak corresponding to one cycle, that does not prove that the data contain deterministic cycles; a stochastic process, such as an AR(2) model with certain coefficient values, can generate a large (fairly broad) peak at the low-frequency end of the spectrum. Alternatively, the fact that an AR(2) model provides a good fit to a time series does not prove that the time series contains stochastic cycles, because even a time series that contains perfect cosine waveforms can be well fitted by an AR(2) model.

Perhaps an analogy will be helpful. ANOVA and regression each carry out similar variance partitioning, but they report and highlight somewhat different features of the data. Similarly, both frequency-domain and time-domain analyses of time-series data basically use the same information, but the results are presented in different ways that highlight different aspects of the results. However, the ability to make a causal inference depends on design, not on whether the researcher happened to analyze her or his data using ANOVA or regression. Similarly, the ability to make inferences about the nature of the process that generates the time series (deterministic vs. stochastic, endogenous vs. exogenous, and causality vs. mere covariation between time series) depends on the research design of the time-series study and on the circumstances under which the data were collected, not on whether the researcher happened to choose frequency-domain or time-domain analyses as a means of data analysis.

Thus, a researcher chooses a frequency-domain or a time-domain approach to data analysis because she or he hopes to highlight certain features of the data. Researchers who are interested in detecting and describing cycles typically use frequency-domain techniques. Researchers who are not interested in cycles (except as a possible source of artifact that needs to be removed prior to further model fitting) tend to use time-domain methods.

Empirical Example: Correspondence of Some Frequency-Domain and Time-Domain Results

Warner (1992b) analyzed the patterns in the amount of vocal activity over time in 110 on–off vocal activity time series using both time-domain methods (fitting a second-order autoregressive time-series model to each time series) and frequency-domain methods (periodogram analysis of each time series). An index of pattern derived from the autoregression (the R^2 describing how well an AR(2) model fit each time series) was correlated +.86 with an index of pattern derived from the frequency-domain analysis (the percentage of variance accounted for by the five largest peaks at the low-frequency end of the periodogram). This suggests that, at least for this particular type of time-series data, these two indexes (autoregressive R^2 and the proportion of variance explained by a few long cycles) contain redundant or interchangeable information. The correspondence between a summary statistic derived from a time-domain analysis and a second summary statistic derived from a frequency-domain analysis should not be expected to be this strong in all cases. (It is possible to have situations in which a large autoregressive R^2 could arise without any cycles in the data.)

The point of this discussion is that sometimes studies of behavioral time series that employ different statistical methods (time domain vs. frequency domain) may be describing the same phenomena in different ways (much as an ANOVA and a regression analysis of the same data set may be highlighting different features while essentially telling the same story about the relations among variables). Equivalence of findings between time-series regression studies and periodogram analysis studies should not be assumed automatically, but it may occur in some situations. Therefore it may well be useful for researchers who prefer to use frequency-domain analyses (such as the ones covered in this book) and the researchers who prefer to use time-domain analyses (such as Box–Jenkins ARIMA modeling and time-series regression) to read each other's results and recognize the possible implications.

Chapter Summary

This book outlines statistical procedures for describing time series that are based on the familiar idea of variance partitioning. For a single time series, the variance is partitioned into variance due to (1) linear or curvilinear trend; (2) one or more relatively regular cyclic components; and (3) a set of residuals from the trends and cycles, which ideally should be random or white noise. When the relationship between a pair of time series is examined, it is possible to focus on relations between any or all of these components. The classic econometric (or Box–Jenkins) approach to time-series analysis has generally involved looking only at relationships between the residuals of the two time series; in this book, it is suggested that the researcher ought also to look at relations between the cycles, or even relations between the trend components, of the two time series—if there is reason to believe that these relations are not merely spurious.

In general, it is suggested that the analyst use the simplest possible analysis (given the research goals). If the goal of the research is primarily descriptive, and if neither causal inference nor significance testing are important, then it may be justifiable to use very simple summary indexes (such as the overall correlation between the two time series) as descriptive information. If the researcher wants to make causal inferences or wants to have the independent residuals that are needed to do unbiased significance tests, then more complicated analyses are needed—for instance, removing the trend and the cycles from one or both time series and then looking at time-lagged relations between the residuals. In addition, if strong causal inferences are desired, researchers may need to take a more experimental approach to the design of time-series studies.

However, at this stage in the development of research on social interaction, relations between mood and physiology over time, and other behavioral/social science time-series research questions, many researchers simply need to obtain useful descriptive information about the patterning of mood, behavior, or physiology over time for individuals or dyads. Summary indexes can be obtained to describe such patterning or coordination (such as the length of the cycle that explains the largest amount of variance; the percentage of variance accounted for by the strongest cyclic component; or the correlation or coherence between time series within a specific frequency band). These indexes of pattern or coordination can then be used as either independent or dependent variables in a larger research context. For example, one can ask how the strength of coordination between infant and caretaker behavior (as indexed by squared coherence, or some simpler statistic such as a lagged cross-correlation) is affected by an experimental intervention; or one can ask whether the strength of coordination predicts later outcomes (such as later measures of infant attachment style or temperament).

Readers who are primarily concerned with either significance testing or causal inference will want to consult other books that present more orthodox treatment of these topics. Several sources are particularly recommended. Gottman (1981b) gives a somewhat more formal treatment of both spectral analysis and time-series analysis in his introduction to time series for social scientists. Ostrom (1978) provides a good brief overview of issues in time-series regression. The present book has sought to provide a clear understanding of the motivation for these more complicated analytic procedures described by other authors—and also a justification for sometimes using less complicated analytic procedures, when this is consistent with the goals of the researcher.

Raw Time-Series Data

This appendix contains the raw time-series data (Tables A.1–A.7) that were used in all the examples presented in the book. They are printed (using the LIST command) from SPSS for Windows data worksheets.

TABLE A.1. *airline.sav*   International Airline Passengers: Monthly Totals (1000's of Passengers), January 1949–December 1960

Number of cases listed: 144

DATE_	PASSENGR	DATE_	PASSENGR	DATE_	PASSENGR
JAN 1949	112.00	JAN 1953	196.00	JAN 1957	315.00
FEB 1949	118.00	FEB 1953	196.00	FEB 1957	301.00
MAR 1949	132.00	MAR 1953	236.00	MAR 1957	356.00
APR 1949	129.00	APR 1953	235.00	APR 1957	348.00
MAY 1949	121.00	MAY 1953	229.00	MAY 1957	355.00
JUN 1949	135.00	JUN 1953	243.00	JUN 1957	422.00
JUL 1949	148.00	JUL 1953	264.00	JUL 1957	465.00
AUG 1949	148.00	AUG 1953	272.00	AUG 1957	467.00
SEP 1949	136.00	SEP 1953	237.00	SEP 1957	404.00
OCT 1949	119.00	OCT 1953	211.00	OCT 1957	347.00
NOV 1949	104.00	NOV 1953	180.00	NOV 1957	305.00
DEC 1949	118.00	DEC 1953	201.00	DEC 1957	336.00
JAN 1950	115.00	JAN 1954	204.00	JAN 1958	340.00
FEB 1950	126.00	FEB 1954	188.00	FEB 1958	318.00
MAR 1950	141.00	MAR 1954	235.00	MAR 1958	362.00
APR 1950	135.00	APR 1954	227.00	APR 1958	348.00
MAY 1950	125.00	MAY 1954	234.00	MAY 1958	363.00
JUN 1950	149.00	JUN 1954	264.00	JUN 1958	435.00
JUL 1950	170.00	JUL 1954	302.00	JUL 1958	491.00
AUG 1950	170.00	AUG 1954	293.00	AUG 1958	505.00
SEP 1950	158.00	SEP 1954	259.00	SEP 1958	404.00
OCT 1950	133.00	OCT 1954	229.00	OCT 1958	359.00
NOV 1950	114.00	NOV 1954	203.00	NOV 1958	310.00
DEC 1950	140.00	DEC 1954	229.00	DEC 1958	337.00
JAN 1951	145.00	JAN 1955	242.00	JAN 1959	360.00
FEB 1951	150.00	FEB 1955	233.00	FEB 1959	343.00
MAR 1951	178.00	MAR 1955	267.00	MAR 1959	406.00
APR 1951	163.00	APR 1955	269.00	APR 1959	396.00
MAY 1951	172.00	MAY 1955	270.00	MAY 1959	420.00
JUN 1951	178.00	JUN 1955	315.00	JUN 1959	472.00
JUL 1951	199.00	JUL 1955	364.00	JUL 1959	548.00
AUG 1951	199.00	AUG 1955	347.00	AUG 1959	559.00
SEP 1951	184.00	SEP 1955	312.00	SEP 1959	463.00
OCT 1951	162.00	OCT 1955	274.00	OCT 1959	407.00
NOV 1951	146.00	NOV 1955	237.00	NOV 1959	362.00
DEC 1951	166.00	DEC 1955	278.00	DEC 1959	405.00
JAN 1952	171.00	JAN 1956	284.00	JAN 1960	417.00
FEB 1952	180.00	FEB 1956	277.00	FEB 1960	391.00
MAR 1952	193.00	MAR 1956	317.00	MAR 1960	419.00
APR 1952	181.00	APR 1956	313.00	APR 1960	461.00
MAY 1952	183.00	MAY 1956	318.00	MAY 1960	472.00
JUN 1952	218.00	JUN 1956	374.00	JUN 1960	535.00
JUL 1952	230.00	JUL 1956	413.00	JUL 1960	622.00
AUG 1952	242.00	AUG 1956	405.00	AUG 1960	606.00
SEP 1952	209.00	SEP 1956	355.00	SEP 1960	508.00
OCT 1952	191.00	OCT 1956	306.00	OCT 1960	461.00
NOV 1952	172.00	NOV 1956	271.00	NOV 1960	390.00
DEC 1952	194.00	DEC 1956	306.00	DEC 1960	432.00

Source: Box and Jenkins (1970, p. 531).

TABLE A.2. *mood.sav* Simulated Daily
Mood Ratings for *N* = 14 Days;
Data Created by Warner

Number of cases listed: 14

DAY	MOOD
0	4.99
1	3.37
2	3.06
3	2.41
4	2.91
5	3.69
6	4.88
7	5.35
8	2.37
9	2.59
10	2.75
11	2.42
12	3.51
13	4.70

TABLE A.3. *sbp.sav* Systolic Blood Pressure Readings, One Mean SBP Value for Each 10-Second Time Interval for 1,280 Seconds; Data Collected by Warner Using the Ohmeda Finapres Finger Cuff Noninvasive Blood Pressure Monitor

Number of cases listed: 128

SECONDS	SBP	SECONDS	SBP	SECONDS	SBP
10	157.60	440	164.20	870	156.60
20	160.80	450	164.00	880	156.20
30	162.40	460	161.60	890	158.20
40	160.60	470	160.60	900	162.20
50	165.00	480	156.20	910	162.60
60	163.00	490	151.00	920	167.00
70	165.00	500	155.80	930	168.60
80	159.80	510	157.00	940	169.60
90	158.40	520	160.40	950	165.20
100	160.80	530	160.00	960	162.60
110	160.60	540	161.60	970	163.00
120	160.60	550	164.40	980	164.40
130	159.80	560	163.40	990	168.00
140	162.00	570	161.40	1000	167.60
150	162.40	580	158.20	1010	162.00
160	167.20	590	162.80	1020	153.00
170	166.80	600	159.00	1030	156.40
180	166.00	610	158.00	1040	166.60
190	170.80	620	162.80	1050	164.20
200	168.00	630	160.60	1060	161.00
210	164.80	640	165.60	1070	164.00
220	166.40	650	160.00	1080	163.40
230	163.60	660	156.00	1090	169.40
240	161.80	670	152.80	1100	172.00
250	167.60	680	154.60	1110	168.40
260	167.40	690	157.80	1120	171.00
270	168.20	700	166.20	1130	167.60
280	165.80	710	165.80	1140	168.60
290	164.00	720	164.80	1150	166.40
300	160.60	730	164.40	1160	169.80
310	160.40	740	163.60	1170	163.60
320	158.20	750	169.40	1180	165.40
330	160.80	760	167.40	1190	161.00
340	164.80	770	163.80	1200	160.00
350	158.60	780	165.00	1210	157.00
360	157.80	790	166.60	1220	158.40
370	157.00	800	158.80	1230	159.80
380	152.60	810	159.60	1240	157.60
390	157.60	820	155.80	1250	157.00
400	168.20	830	156.20	1260	155.20
410	164.80	840	157.80	1270	158.80
420	169.40	850	158.20	1280	164.80
430	163.00	860	154.00		

TABLE A.4. *simtalk.sav*   Simulated Data Used to Illustrate Sources of Artifact
in Frequency Analysis; Data Created by Warner

Number of cases listed: 150

T	SIMTALK	SIMTRND	SIMSPK	BOXCAR	SIMCYC
0	.39	.00	.30	.00	.91
1	.56	.08	.56	.00	.95
2	.51	.07	.51	.00	.87
3	.49	.06	.49	.00	.78
4	.21	.00	.21	.00	.56
5	.23	.00	.23	.00	.47
6	.50	.08	.50	.00	.50
7	.87	.27	.87	.00	.60
8	.53	.10	.53	.00	.47
9	.50	.10	.50	.00	.50
10	.50	.10	.50	.00	.57
11	.34	.03	.34	.00	.61
12	.83	.28	.83	.00	.91
13	.40	.06	.40	.00	.83
14	.48	.11	.48	.00	.92
15	.23	.00	.23	.23	.84
16	.28	.02	.28	.28	.84
17	.75	.26	.75	.75	.97
18	.70	.24	5.70	.70	.86
19	.67	.23	.67	.67	.74
20	.57	.18	.57	.57	.60
21	.40	.10	.40	.40	.46
22	.85	.33	.85	.85	.59
23	.51	.17	.51	.51	.46
24	.56	.20	.56	.56	.52
25	.82	.33	.82	.82	.70
26	.52	.19	.52	.52	.68
27	.50	.19	.50	.50	.78
28	.69	.29	.69	.69	.94
29	.38	.13	.38	.38	.88
30	.56	.23	.56	.56	.97
31	.76	.34	.76	.76	1.03
32	.30	.11	.30	.30	.79
33	.51	.22	5.51	.51	.78
34	.42	.18	.42	.42	.64
35	.50	.23	.50	.50	.58
36	.35	.15	.35	.35	.44
37	.78	.37	.78	.78	.57
38	.82	.40	.82	.82	.58
39	.33	.16	.33	.33	.43
40	.56	.28	.56	.56	.60
41	.64	.32	.64	.64	.73
42	.57	.29	.57	.57	.80
43	.64	.33	.64	.64	.92
44	.59	.31	.59	.59	.96
45	.50	.28	.50	.50	.95
46	.72	.39	.72	.72	1.02
47	.57	.32	.57	.57	.90
48	.57	.33	.57	.57	.81
49	.62	.36	.62	.62	.72

(*continued*)

TABLE A.4. (*continued*)

T	SIMTALK	SIMTRND	SIMSPK	BOXCAR	SIMCYC
50	.34	.22	.34	.34	.51
51	.05	.08	.05	.05	.32
52	.71	.42	.71	.71	.54
53	.83	.48	.83	.83	.59
54	.76	.45	.76	.76	.60
55	.64	.39	.64	.64	.63
56	.38	.27	.38	.38	.63
57	.48	.32	.48	.48	.77
58	1.00	.59	1.00	1.00	1.07
59	.44	.31	.44	.44	.90
60	.47	.34	.47	.47	.94
61	.76	.49	.76	.76	1.03
62	.74	.48	.74	.74	.97
63	.35	.29	.35	.35	.72
64	.89	.57	.89	.89	.83
65	.65	.45	.65	.65	.64
66	.44	.35	.44	.44	.47
67	.54	.41	.54	.54	.47
68	.78	.53	.78	.78	.57
69	.76	.52	.76	.76	.60
70	.16	.23	.16	.16	.44
71	.88	.60	.88	.88	.83
72	.48	.40	.48	.48	.77
73	.65	.49	.65	.65	.93
74	.55	.44	.55	.55	.95
75	.30	.32	.30	.30	.87
76	.73	.54	.73	.73	1.02
77	.47	.42	.47	.47	.86
78	.39	.39	.39	.39	.74
79	.19	.29	.19	.19	.55
80	.71	.55	.71	.71	.66
81	.48	.45	.48	.48	.49
82	.41	.41	.41	.41	.42
83	.40	.42	.40	.40	.42
84	.18	.31	.18	.18	.37
85	.47	.46	.47	.47	.56
86	.19	.33	.19	.19	.55
87	.24	.35	.24	.24	.67
88	.63	.56	.63	.63	.92
89	.70	.59	.70	.70	1.01
90	.46	.48	.46	.46	.93
91	.63	.57	.63	.63	.98
92	.88	.70	.88	.88	1.02
93	.58	.55	.58	.58	.81
94	.70	.62	.70	.70	.76
95	.31	.43	.31	.31	.50
96	.37	.46	.37	.37	.45
97	.37	.47	.37	.37	.40
98	.52	.55	.52	.52	.46
99	.61	.60	.61	.61	.54

TABLE A.4. (*continued*)

T	SIMTALK	SIMTRND	SIMSPK	BOXCAR	SIMCYC
100	.33	.46	.33	.33	.50
101	.93	.77	.93	.93	.84
102	.82	.72	.82	.82	.91
103	.43	.53	.43	.43	.84
104	.32	.48	.32	.32	.86
105	.64	.65	.64	.64	1.01
106	.61	.63	.61	.61	.97
107	.37	.52	.37	.37	.82
108	.76	.72	.76	.76	.88
109	.53	.61	.53	.53	.69
110	.46	.58	.46	.46	.56
111	.83	.77	.83	.83	.63
112	.69	.71	.69	.69	.53
113	.57	.65	.57	.57	.48
114	.72	.73	.72	.72	.59
115	.85	.80	.85	.85	.71
116	.52	.64	.52	.52	.68
117	.26	.51	.26	.26	.68
118	.45	.62	.45	.45	.85
119	.71	.75	.71	.71	1.01
120	.47	.64	.47	.47	.94
121	.27	.54	.27	.27	.84
122	.47	.65	.47	.47	.86
123	.60	.72	.60	.60	.82
124	.33	.58	.33	.33	.61
125	.47	.66	.47	.47	.56
126	.64	.75	.64	.64	.56
127	.52	.70	.52	.52	.46
128	.61	.75	.61	.61	.50
129	.66	.78	.66	.66	.56
130	.12	.51	.12	.12	.42
131	.52	.71	.52	.52	.68
132	.49	.71	.49	.49	.77
133	.61	.77	.61	.61	.91
134	.59	.76	.59	.59	.96
135	.73	.84	.73	.00	1.04
136	.27	.61	.27	.00	.84
137	.53	.75	.53	.00	.88
138	.62	.80	.62	.00	.83
139	.64	.81	.64	.00	.73
140	.82	.91	.82	.00	.71
141	.35	.68	.35	.00	.44
142	.67	.84	.67	.00	.52
143	.60	.82	.60	.00	.50
144	.57	.80	.57	.00	.52
145	.85	.95	.85	.00	.71
146	.54	.77	.54	.00	.69
147	.47	.77	.47	.00	.77
148	.28	.68	.28	.00	.78
149	.63	.86	.63	.00	.98

TABLE A.5. *sunspot.sav*   Wolfer Sunspot Data: Numbers of Sunspots Observed Yearly, for the Years 1770–1869

Number of cases listed: 100

YEAR	SUNSPOT	YEAR	SUNSPOT
1770	101	1820	16
1771	82	1821	7
1772	66	1822	4
1773	35	1823	2
1774	31	1824	8
1775	7	1825	17
1776	20	1826	36
1777	92	1827	50
1778	154	1828	62
1779	125	1829	67
1780	85	1830	71
1781	68	1831	48
1782	38	1832	8
1783	23	1833	8
1784	10	1834	13
1785	24	1835	57
1786	83	1836	122
1787	132	1837	138
1788	131	1838	103
1789	118	1839	86
1790	90	1840	63
1791	67	1841	37
1792	60	1842	24
1793	47	1843	11
1794	41	1844	15
1795	21	1845	40
1796	16	1846	62
1797	6	1847	98
1798	4	1848	124
1799	7	1849	96
1800	14	1850	66
1801	34	1851	64
1802	45	1852	54
1803	43	1853	39
1804	48	1854	21
1805	42	1855	7
1806	28	1856	4
1807	10	1857	23
1808	8	1858	55
1809	2	1859	94
1810	0	1860	96
1811	1	1861	77
1812	5	1862	59
1813	12	1863	44
1814	14	1864	47
1815	35	1865	30
1816	46	1866	16
1817	41	1867	7
1818	30	1868	37
1819	24	1869	74

Source: Box and Jenkins (1970, p. 530).

TABLE A.6. *talkhr.sav* Data on Amount of Talk and Heart Rate, Each Measured Once Every 2 Seconds, for an Individual Speaker in a Conversation; Data Collected by Warner

Number of cases listed: 240

SECONDS	FTALK	FHR	SECONDS	FTALK	FHR	SECONDS	FTALK	FHR
2	.88	70	102	1.00	74	202	.50	74
4	.63	63	104	.88	74	204	.00	75
6	.38	63	106	.88	70	206	.38	75
8	1.00	64	108	1.00	70	208	.88	64
10	1.00	58	110	1.00	56	210	1.00	62
12	1.00	58	112	.13	56	212	.50	56
14	.88	58	114	.38	56	214	.00	73
16	1.00	58	116	.13	56	216	.00	73
18	.25	70	118	.25	61	218	.00	65
20	.25	75	120	.00	57	220	.13	50
22	.13	78	122	.38	57	222	.25	53
24	.50	78	124	.38	60	224	.88	69
26	.25	78	126	.13	55	226	.75	69
28	1.00	72	128	.88	69	228	.00	61
30	.88	72	130	1.00	69	230	.75	53
32	.13	59	132	1.00	62	232	.38	56
34	.63	59	134	1.00	63	234	.88	48
36	.38	56	136	1.00	62	236	.50	48
38	.38	57	138	.88	62	238	1.00	48
40	.25	61	140	.75	65	240	1.00	59
42	.00	61	142	.13	59	242	1.00	58
44	.25	50	144	.50	59	244	1.00	56
46	.75	52	146	1.00	59	246	1.00	56
48	.13	52	148	.88	59	248	1.00	65
50	1.00	68	150	.75	65	250	1.00	68
52	.25	68	152	1.00	65	252	1.00	71
54	.00	55	154	1.00	65	254	.75	71
56	.25	62	156	1.00	69	256	.38	75
58	.88	67	158	1.00	69	258	.00	70
60	.88	67	160	1.00	74	260	.00	63
62	.00	66	162	1.00	72	262	.25	63
64	.63	68	164	1.00	72	264	.38	65
66	.13	68	166	.88	68	266	.25	62
68	.00	59	168	1.00	68	268	1.00	62
70	.00	59	170	1.00	61	270	1.00	61
72	.00	57	172	1.00	56	272	.88	60
74	.63	56	174	1.00	56	274	1.00	66
76	.00	57	176	1.00	65	276	1.00	57
78	.25	57	178	.88	62	278	.88	65
80	1.00	57	180	.63	62	280	1.00	65
82	1.00	57	182	.00	69	282	1.00	67
84	1.00	57	184	.00	69	284	.88	60
86	.88	55	186	.00	69	286	1.00	64
88	1.00	55	188	.50	69	288	1.00	68
90	1.00	64	190	.38	50	290	1.00	68
92	1.00	59	192	1.00	67	292	.88	74
94	1.00	59	194	1.00	57	294	1.00	74
96	1.00	59	196	1.00	57	296	1.00	74
98	1.00	69	198	1.00	66	298	1.00	74
100	1.00	72	200	.88	70	300	1.00	59

(continued)

TABLE A.6. (*continued*)

SECONDS	FTALK	FHR	SECONDS	FTALK	FHR
302	1.00	66	392	.50	52
304	.88	75	394	1.00	53
306	1.00	70	396	1.00	53
308	1.00	70	398	1.00	55
310	1.00	71	400	1.00	64
312	.88	71	402	1.00	64
314	1.00	71	404	1.00	69
316	1.00	71	406	.63	62
318	1.00	71	408	1.00	75
320	1.00	73	410	.38	70
322	.75	73	412	.50	70
324	1.00	63	414	.00	77
326	.25	63	416	.00	70
328	.00	65	418	.00	59
330	.38	48	420	.00	54
332	1.00	59	422	.00	54
334	.88	54	424	.00	57
336	1.00	54	426	.00	59
338	.88	59	428	.13	52
340	.88	59	430	.50	50
342	1.00	62	432	.38	58
344	.88	58	434	.13	58
346	1.00	58	436	.50	58
348	1.00	58	438	.00	58
350	.88	52	440	.25	58
352	1.00	60	442	.13	56
354	1.00	71	444	.75	57
356	1.00	71	446	.00	56
358	.75	62	448	.38	56
360	1.00	67	450	.00	56
362	1.00	66	452	.00	64
364	1.00	66	454	.00	59
366	1.00	70	456	.00	60
368	1.00	70	458	.50	53
370	1.00	71	460	.13	56
372	1.00	70	462	.00	56
374	.75	70	464	.00	55
376	.13	73	466	.38	58
378	.00	73	468	.00	55
380	1.00	59	470	.00	55
382	.75	52	472	.00	63
384	.25	52	474	.00	54
386	.25	52	476	.63	57
388	.00	60	478	.25	54
390	.00	60	480	.00	66

TABLE A.7. *talkmf.sav* Data on Amount of Talk Measured Once Every 10
Seconds for Each of the Two Speakers in a Dyadic Conversation;
Data Collected by Warner

Number of cases listed: 150

SECONDS	MTALK	FTALK	SECONDS	MTALK	FTALK	SECONDS	MTALK	FTALK
10	.28	.93	510	.00	.68	1010	.00	.78
20	.10	.73	520	.43	.55	1020	.00	.90
30	.33	.75	530	.53	.05	1030	.00	.98
40	.23	.73	540	.85	.15	1040	.03	.75
50	.25	.75	550	.78	.03	1050	.00	.88
60	.43	.75	560	.73	.05	1060	.03	.95
70	.50	.88	570	.70	.03	1070	.28	.78
80	.35	.83	580	.65	.05	1080	.65	.88
90	.33	1.00	590	.68	.50	1090	.48	.55
100	.43	.57	600	.60	.20	1100	.13	.85
110	.63	.65	610	.15	.95	1110	.00	.85
120	.25	.73	620	.08	.88	1120	.03	.80
130	.57	.45	630	.40	.57	1130	.15	.50
140	.88	.03	640	.33	.83	1140	.03	.73
150	.70	.18	650	.00	.83	1150	.30	.50
160	.88	.20	660	.30	.85	1160	.05	.75
170	.80	.05	670	.00	.85	1170	.05	.75
180	.45	.53	680	.13	.90	1180	.30	.40
190	.00	.88	690	.00	.63	1190	.88	.15
200	.25	.95	700	.03	.93	1200	.53	.20
210	.08	80	710	.23	.85	1210	.80	.18
220	.10	.78	720	.05	.85	1220	.63	.03
230	.53	.57	730	.13	.57	1230	.65	.08
240	.23	.80	740	.05	.68	1240	.88	.08
250	.33	.38	750	.00	.83	1250	.78	.05
260	.08	.65	760	.18	.60	1260	.78	.13
270	.13	.55	770	.00	.73	1270	.08	.88
280	.00	.68	780	.03	.85	1280	.13	.83
290	.00	.88	790	.00	.63	1290	.53	.45
300	.00	.95	800	.33	.13	1300	.68	.60
310	.00	.80	810	.55	.30	1310	.35	.83
320	.08	1.00	820	.75	.03	1320	.50	.28
330	.05	.95	830	.73	.35	1330	.33	.48
340	.08	.83	840	.73	.10	1340	.83	.45
350	.10	.60	850	.85	.03	1350	.45	.15
360	.23	.88	860	.83	.03	1360	.73	.33
370	.13	.78	870	.88	.10	1370	.00	.70
380	.28	.88	880	.73	.00	1380	.00	.85
390	.23	.85	890	.70	.05	1390	.18	.90
400	.05	.78	900	.80	.08	1400	.33	.95
410	.25	.85	910	.93	.10	1410	.25	.78
420	.25	.83	920	.63	.03	1420	.18	.95
430	.03	.78	930	.85	.23	1430	.35	1.00
440	.00	.65	940	.15	.75	1440	.05	.98
450	.25	.80	950	.10	.83	1450	.38	.60
460	.03	.95	960	.30	.57	1460	.80	.15
470	.25	.73	970	.53	.35	1470	.08	.65
480	.50	.50	980	.83	.15	1480	.40	.68
490	.18	.78	990	.40	.33	1490	.28	.63
500	.20	.80	1000	.00	.80	1500	.18	.88

Critical Values for the Fisher Test of Significance for Periodogram Analysis

The critical values presented below in Tables B.1–B.3 were computed from a formula given by Russell (1985).

To compute the test statistic g, compute the periodogram for the time series. Find the sum of all the periodogram ordinates, and then divide the periodogram intensity for each frequency by this sum to yield an estimate of the proportion of variance in the time series that is accounted for by each frequency component represented in the periodogram. The test statistic g is simply the proportion of the total variance in the time series that is accounted for by a particular frequency in the periodogram analysis.

Select the largest values of this proportion, that is, the proportions of variance that are explained by the first largest, second largest, third largest, fourth largest, and fifth largest periodogram ordinates. The tables provided here give critical values to test the significance of the peaks that are ranked first through fifth largest in the proportion of variance accounted for.

To choose the appropriate critical values, you need to know N (number of observations in the time series), and the alpha level for your significance tests. Critical values of g are given for $\alpha = .05$, $\alpha = .01$, and $\alpha = .001$ in separate tables. Within the table, the columns headed r1, r2, r3, r4, and r5 give the critical values used to test the significance of the peaks that are ranked first, second, . . . , fifth in the periodogram.

Start with the largest obtained proportion of variance and compare this to the critical value in the r1 column (critical value of g for the largest periodogram ordinate). If the obtained proportion of variance exceeds the critical value given in the table then this first peak is statistically significant. Smaller peaks may be tested only if the larger peaks were significant.

For example, for a time series with $N = 100$, suppose that the proportion of variance accounted for by the five largest peaks looked like this:

Rank of periodogram ordinate	1st	2nd	3rd	4th	5th
Obtained percentage of variance	.2100	.1267	.0654	.0603	.0333
Critical values from table (with $N = 100$, $\alpha = .05$)	.1335	.0938	.0769	.0666	.0593

The first and second largest peaks would be judged significant; the third, fourth, and fifth largest peaks would be judged nonsignificant.

TABLE B.1. Critical Values of the Proportion of Variance for the Fisher Test, $\alpha = .05$

n	r1	r2	r3	r4	r5
25	.41688	.25166	.18464	.14541	.11833
30	.35172	.21905	.16449	.13226	.11004
35	.31923	.20204	.15351	.12472	.10482
40	.28104	.18136	.13978	.11496	.09775
45	.26061	.16999	.13204	.10932	.09353
50	.23534	.15561	.12207	.10191	.08786
55	.22123	.14742	.11630	.09756	.08447
60	.20318	.13678	.10870	.09174	.07988
65	.19281	.13058	.10422	.08828	.07711
70	.17922	.12236	.09822	.08359	.07333
75	.17125	.11748	.09463	.08077	.07103
80	.16062	.11092	.08976	.07690	.06786
85	.15429	.10697	.08681	.07455	.06592
90	.14574	.10160	.08277	.07131	.06323
95	.14058	.09833	.08030	.06931	.06157
100	.13354	.09384	.07689	.06655	.05925
105	.12924	.09109	.07478	.06483	.05781
110	.12334	.08728	.07186	.06244	.05579
115	.11971	.08493	.07004	.06095	.05453
120	.11467	.08165	.06751	.05886	.05275
125	.11156	.07961	.06592	.05756	.05164
130	.10722	.07675	.06370	.05571	.05006
135	.10452	.07497	.06231	.05456	.04907
140	.10073	.07246	.06034	.05292	.04766
145	.09836	.07089	.05910	.05188	.04677
150	.09503	.06866	.05734	.05041	.04550
155	.09293	.06726	.05624	.04948	.04470
160	.08997	.06527	.05466	.04816	.04355
165	.08811	.06401	.05366	.04732	.04282
170	.08546	.06222	.05224	.04612	.04178
175	.08379	.06109	.05134	.04536	.04111
180	.08141	.05947	.05004	.04426	.04016

TABLE B.1. (*continued*)

n	r1	r2	r3	r4	r5
185	.07990	.05844	.04922	.04356	.03955
190	.07774	.05697	.04804	.04256	.03867
195	.07637	.05603	.04729	.04192	.03811
200	.07441	.05469	.04621	.04100	.03730
205	.07316	.05383	.04551	.04041	.03679
210	.07137	.05259	.04452	.03956	.03604
215	.07023	.05181	.04388	.03901	.03556
220	.06859	.05067	.04296	.03823	.03486
225	.06754	.04994	.04237	.03772	.03442
230	.06602	.04889	.04152	.03699	.03377
235	.06505	.04822	.04097	.03652	.03336
240	.06365	.04724	.04018	.03584	.03275
245	.06276	.04661	.03967	.03540	.03237
250	.06146	.04571	.03893	.03476	.03180
255	.06063	.04513	.03845	.03435	.03144
260	.05942	.04428	.03776	.03376	.03091
265	.05864	.04373	.03732	.03337	.03057
270	.05752	.04294	.03667	.03281	.03007
275	.05679	.04243	.03625	.03245	.02975
280	.05574	.04169	.03564	.03192	.02928
285	.05507	.04121	.03525	.03158	.02897
290	.05408	.04052	.03468	.03109	.02853
295	.05344	.04007	.03431	.03076	.02825
300	.05252	.03941	.03377	.03030	.02783
305	.05192	.03899	.03342	.02999	.02756
310	.05105	.03837	.03291	.02955	.02716
315	.05049	.03797	.03258	.02926	.02690
320	.04967	.03739	.03210	.02884	.02653
325	.04914	.03701	.03178	.02857	.02628
330	.04836	.03645	.03133	.02817	.02592
335	.04786	.03610	.03103	.02791	.02569
340	.04713	.03557	.03059	.02753	.02535

(*continued*)

TABLE B.1. (*continued*)

n	r1	r2	r3	r4	r5
345	.04665	.03523	.03031	.02729	.02513
350	.04596	.03473	.02990	.02692	.02481
355	.04550	.03441	.02963	.02669	.02460
360	.04485	.03394	.02924	.02635	.02429
365	.04442	.03363	.02898	.02612	.02409
370	.04379	.03318	.02861	.02579	.02379
375	.04338	.03289	.02836	.02558	.02360
380	.04279	.03246	.02801	.02527	.02332
385	.04240	.03218	.02777	.02506	.02313
390	.04183	.03177	.02743	.02476	.02286
395	.04146	.03150	.02721	.02457	.02269
400	.04092	.03111	.02688	.02428	.02243
405	.04057	.03086	.02667	.02409	.02226
410	.04005	.03048	.02636	.02382	.02201
415	.03971	.03024	.02615	.02364	.02185
420	.03922	.02988	.02585	.02337	.02161
425	.03889	.02964	.02566	.02320	.02145
430	.03842	.02930	.02537	.02295	.02123
435	.03811	.02908	.02518	.02278	.02108
440	.03766	.02875	.02490	.02254	.02086
445	.03736	.02853	.02472	.02238	.02071
450	.03693	.02821	.02446	.02215	.02050
455	.03664	.02801	.02428	.02199	.02036
460	.03622	.02770	.02403	.02177	.02016
465	.03595	.02750	.02386	.02162	.02002
470	.03555	.02721	.02362	.02140	.01983
475	.03529	.02702	.02346	.02126	.01970
480	.03490	.02674	.02322	.02105	.01951
485	.03465	.02655	.02306	.02091	.01938
490	.03428	.02628	.02283	.02071	.01920
495	.03403	.02610	.02268	.02058	.01908
500	.03368	.02584	.02246	.02038	.01890

TABLE B.2. Critical Values of g for the Fisher Test, $\alpha = .01$

n	r1	r2	r3	r4	r5
25	.50357	.28869	.20709	.16157	.13140
30	.42722	.25200	.18465	.14668	.12142
35	.38851	.23266	.17238	.13822	.11541
40	.34258	.20900	.15699	.12732	.10741
45	.31784	.19593	.14829	.12102	.10267
50	.28709	.17935	.13707	.11276	.09634
55	.26987	.16989	.13057	.10790	.09257
60	.24778	.15758	.12200	.10143	.08748
65	.23507	.15039	.11694	.09757	.08441
70	.21840	.14086	.11016	.09235	.08023
75	.20860	.13520	.10610	.08920	.07768
80	.19555	.12758	.10060	.08490	.07419
85	.18776	.12299	.09727	.08228	.07204
90	.17724	.11675	.09270	.07867	.06907
95	.17090	.11296	.08990	.07645	.06724
100	.16223	.10774	.08604	.07337	.06468
105	.15695	.10455	.08367	.07146	.06309
110	.14968	.10012	.08036	.06880	.06087
115	.14522	.09739	.07831	.06714	.05948
120	.13902	.09358	.07544	.06482	.05752
125	.13520	.09122	.07365	.06337	.05629
130	.12986	.08790	.07114	.06132	.05456
135	.12654	.08583	.06957	.06003	.05347
140	.12188	.08292	.06734	.05821	.05192
145	.11898	.08109	.06594	.05706	.05094
150	.11488	.07851	.06396	.05542	.04954
155	.11231	.07688	.06271	.05439	.04866
160	.10867	.07458	.06093	.05292	.04740
165	.10639	.07312	.05981	.05199	.04659
170	.10314	.07105	.05820	.05065	.04545
175	.10109	.06973	.05718	.04980	.04472
180	.09817	.06786	.05572	.04859	.04367

(*continued*)

TABLE B.2. (*continued*)

n	rl	r2	r3	r4	r5
185	.09632	.06667	.05480	.04781	.04300
190	.09367	.06496	.05347	.04670	.04203
195	.09200	.06388	.05262	.04599	.04142
200	.08960	.06232	.05140	.04497	.04053
205	.08807	.06133	.05062	.04431	.03996
210	.08587	.05990	.04950	.04337	.03914
215	.08448	.05899	.04878	.04277	.03861
220	.08247	.05768	.04774	.04189	.03785
225	.08118	.05683	.04708	.04133	.03736
230	.07933	.05562	.04612	.04052	.03665
235	.07814	.05484	.04550	.04000	.03620
240	.07644	.05372	.04461	.03925	.03554
245	.07534	.05299	.04404	.03876	.03511
250	.07376	.05195	.04321	.03805	.03449
255	.07274	.05127	.04267	.03760	.03409
260	.07127	.05030	.04189	.03694	.03351
265	.07032	.04967	.04139	.03651	.03314
270	.06895	.04876	.04066	.03589	.03259
275	.06806	.04817	.04019	.03549	.03224
280	.06678	.04732	.03951	.03491	.03173
285	.06595	.04677	.03907	.03453	.03139
290	.06475	.04596	.03843	.03398	.03091
295	.06398	.04544	.03801	.03363	.03060
300	.06285	.04469	.03741	.03311	.03014
305	.06212	.04420	.03701	.03277	.02984
310	.06106	.04349	.03644	.03228	.02941
315	.06038	.04303	.03607	.03197	.02913
320	.05938	.04236	.03553	.03150	.02871
325	.05873	.04192	.03518	.03120	.02845
330	.05779	.04129	.03467	.03076	.02806
335	.05718	.04087	.03433	.03047	.02780
340	.05629	.04027	.03385	.03005	.02743

TABLE B.2. (*continued*)

n	rl	r2	r3	r4	r5
345	.05571	.03988	.03353	.02978	.02719
350	.05487	.03931	.03307	.02938	.02683
355	.05432	.03894	.03277	.02912	.02660
360	.05352	.03839	.03233	.02875	.02627
365	.05300	.03804	.03204	.02850	.02605
370	.05224	.03752	.03162	.02814	.02572
375	.05174	.03719	.03135	.02790	.02551
380	.05102	.03669	.03095	.02755	.02520
385	.05055	.03637	.03069	.02733	.02500
390	.04986	.03590	.03030	.02700	.02471
395	.04941	.03560	.03005	.02678	.02452
400	.04875	.03515	.02969	.02647	.02423
405	.04833	.03485	.02945	.02626	.02405
410	.04770	.03442	.02910	.02596	.02378
415	.04729	.03414	.02887	.02576	.02360
420	.04669	.03373	.02854	.02547	.02334
425	.04630	.03347	.02832	.02528	.02317
430	.04573	.03307	.02799	.02500	.02292
435	.04535	.03281	.02778	.02481	.02276
440	.04480	.03244	.02747	.02455	.02252
445	.04444	.03219	.02727	.02437	.02236
450	.04392	.03183	.02698	.02411	.02213
455	.04357	.03159	.02678	.02394	.02198
460	.04307	.03124	.02650	.02369	.02175
465	.04274	.03101	.02631	.02353	.02161
470	.04225	.03068	.02603	.02329	.02139
475	.04193	.03046	.02585	.02314	.02125
480	.04147	.03014	.02559	.02291	.02105
485	.04116	.02993	.02542	.02275	.02091
490	.04071	.02961	.02516	.02253	.02071
495	.04042	.02941	.02499	.02239	.02058
500	.03999	.02911	.02475	.02217	.02039

Appendix B

TABLE B.3. Critical Values of *g* for the Fisher Test, α = .001

n	r1	r2	r3	r4	r5
25	.60567	.33215	.23308	.17984	.14572
30	.52020	.29227	.20886	.16362	.13446
35	.47553	.27073	.19540	.15435	.12777
40	.42152	.24400	.17833	.14233	.11890
45	.39203	.22907	.16861	.13536	.11366
50	.35501	.20999	.15600	.12618	.10665
55	.33412	.19905	.14867	.12078	.10247
60	.30717	.18474	.13896	.11355	.09683
65	.29159	.17636	.13322	.10924	.09342
70	.27108	.16521	.12551	.10340	.08878
75	.25900	.15858	.12089	.09987	.08596
80	.24285	.14963	.11461	.09504	.08208
85	.23321	.14424	.11080	.09210	.07969
90	.22015	.13690	.10558	.08804	.07639
95	.21226	.13243	.10238	.08554	.07435
100	.20148	.12629	.09796	.08208	.07151
105	.19491	.12252	.09524	.07993	.06975
110	.18584	.11730	.09145	.07694	.06727
115	.18028	.11407	.08910	.07507	.06572
120	.17255	.10957	.08581	.07246	.06355
125	.16777	.10678	.08376	.07082	.06218
130	.16110	.10286	.08088	.06851	.06025
135	.15695	.10042	.07907	.06706	.05903
140	.15114	.09698	.07652	.06500	.05731
145	.14750	.09482	.07492	.06371	.05622
150	.14238	.09177	.07265	.06187	.05466
155	.13916	.08985	.07121	.06071	.05368
160	.13462	.08712	.06917	.05905	.05228
165	.13176	.08540	.06788	.05800	.05138
170	.12769	.08295	.06604	.05650	.05011
175	.12513	.08140	.06487	.05554	.04929
180	.12147	.07918	.06319	.05417	.04813

TABLE B.3. (*continued*)

n	rI	r2	r3	r4	r5
185	.11916	.07778	.06213	.05330	.04738
190	.11585	.07576	.06060	.05205	.04631
195	.11376	.07448	.05963	.05124	.04562
200	.11075	.07264	.05823	.05009	.04464
205	.10884	.07147	.05734	.04936	.04400
210	.10610	.06979	.05605	.04830	.04309
215	.10435	.06871	.05523	.04761	.04251
220	.10184	.06716	.05404	.04663	.04166
225	.10023	.06617	.05328	.04600	.04112
230	.09792	.06473	.05218	.04509	.04033
235	.09644	.06381	.05148	.04450	.03982
240	.09430	.06249	.05045	.04365	.03909
245	.09293	.06163	.04980	.04310	.03861
250	.09095	.06040	.04884	.04231	.03792
255	.08968	.05961	.04823	.04180	.03748
260	.08784	.05846	.04734	.04106	.03684
265	.08666	.05772	.04677	.04058	.03642
270	.08494	.05664	.04594	.03988	.03581
275	.08384	.05595	.04540	.03943	.03542
280	.08224	.05494	.04462	.03878	.03485
285	.08121	.05429	.04411	.03835	.03448
290	.07971	.05335	.04338	.03773	.03394
295	.07874	.05274	.04290	.03734	.03360
300	.07733	.05185	.04221	.03675	.03309
305	.07643	.05127	.04176	.03638	.03276
310	.07510	.05044	.04111	.03582	.03228
315	.07425	.04989	.04068	.03547	.03196
320	.07300	.04910	.04006	.03495	.03151
325	.07220	.04859	.03966	.03461	.03121
330	.07102	.04784	.03908	.03411	.03078
335	.07026	.04736	.03869	.03379	.03049
340	.06915	.04665	.03814	.03332	.03008

(*continued*)

TABLE B.3. (*continued*)

n	rl	r2	r3	r4	r5
345	.06843	.04619	.03778	.03302	.02981
350	.06738	.04552	.03725	.03257	.02942
355	.06669	.04508	.03691	.03228	.02917
360	.06570	.04444	.03640	.03185	.02879
365	.06505	.04403	.03608	.03157	.02855
370	.06410	.04342	.03560	.03117	.02819
375	.06349	.04303	.03529	.03090	.02796
380	.06259	.04245	.03483	.03052	.02761
385	.06200	.04207	.03453	.03026	.02739
390	.06114	.04152	.03409	.02989	.02707
395	.06058	.04116	.03381	.02965	.02685
400	.05977	.04063	.03339	.02930	.02654
405	.05923	.04029	.03312	.02906	.02633
410	.05845	.03978	.03272	.02872	.02603
415	.05794	.03945	.03246	.02850	.02584
420	.05720	.03897	.03208	.02818	.02555
425	.05671	.03866	.03183	.02796	.02536
430	.05600	.03820	.03146	.02765	.02508
435	.05553	.03789	.03122	.02745	.02490
440	.05485	.03745	.03087	.02714	.02464
445	.05440	.03716	.03064	.02695	.02446
450	.05375	.03674	.03030	.02666	.02421
455	.05332	.03646	.03008	.02647	.02404
460	.05269	.03605	.02976	.02619	.02379
465	.05228	.03578	.02954	.02601	.02363
470	.05168	.03539	.02923	.02574	.02339
475	.05129	.03513	.02903	.02557	.02324
480	.05071	.03476	.02872	.02531	.02301
485	.05033	.03451	.02853	.02514	.02286
490	.04977	.03415	.02824	.02489	.02264
495	.04941	.03391	.02805	.02473	.02249
500	.04887	.03356	.02777	.02449	.02228

References

Aschoff, J. (Ed.). (1981). *Handbook of behavioral neurobiology: Vol. 4. Biological rhythms*. New York: Plenum.

Babkoff, H., Caspy, T., Mikulincer, M., & Sing, H. (1991). Monotonic and rhythmic influences: A challenge for sleep deprivation research. *Psychological Bulletin, 109*, 411–428.

Bakeman, R., & Brown, J. V. (1977). Behavioral dialogues: An approach to the assessment of mother–infant interactions. *Child Development, 48*, 195–203.

Benton, L. A., & Yates, F. E. (1990). Ultradian adrenocortical and circulatory oscillations in conscious dogs. *American Journal of Physiology, 258*(3, Pt. 2), R578–R590.

Berger, C. R., & Bradac, J. J. (1982). *Language and social knowledge: Uncertainty in interpersonal relations*. London: Edward Arnold.

Bloomfield, P. (1976). *Fourier analysis of time series: An introduction*. New York: Wiley.

Box, G. E. P., & Jenkins, G. M. (1970). *Time series analysis: Forecasting and control*. San Francisco: Holden-Day.

Broadbent, H. A., & Maksik, Y. A. (1992). Analysis of periodic data using Walsh functions. *Behavior Research Methods Instruments and Computers, 24*, 238–247.

Buder, E. H. (1991). A non-linear dynamic model of social interaction. *Communication Research, 18*, 174–198.

Buder, E. H. (1996). Dynamics of speech processes in dyadic interactions. In J. H. Watt & C. A. VanLear (Eds.), *Dynamic patterns in communication processes*. Thousand Oaks, CA: Sage.

Campbell, D. T., & Stanley, J. C. (1966). *Experimental and quasi-experimental designs for research*. Chicago: Rand McNally.

Cappella, J. N. (1981). Mutual influence in expressive behavior: Adult–adult and infant–adult dyadic interaction. *Psychological Bulletin, 89*, 101–132.

Cappella, J. N. (1988). Personal relationships, social relationships, and patterns of interaction. In S. W. Duck (Ed.), *Handbook of personal relationships*. New York: Wiley.

Cappella, J. N. (1996). Dynamic coordination of vocal and kinesic behavior in dyadic interaction: Methods, problems, and interpersonal outcomes. In J.

H. Watt & C. A. VanLear (Eds.), *Dynamic patterns in communication processes*. Thousand Oaks, CA: Sage.

Chapple, E. D. (1970). *Culture and biological man: Explorations in behavioral anthropology*. New York: Holt, Rinehart & Winston.

Chatfield, C. (1991) *The analysis of time series: Theory and practice*. London: Chapman & Hall.

Cook, J., Tyson, R., White, J., Rushe, R. Gottman, J. M., & Murray, J. (1995). Mathematics of marital conflict: Qualitative dynamic mathematical modeling of marital interaction. [Special section: Methodological advances in family psychology.] *Journal of Family Psychology, 9*, 110–130.

Dukes, W. F. (1965). *N* = 1. *Psychological Bulletin, 64*, 74–79.

Field, T. (1985). Attachment as psychobiological attunement: Being on the same wavelength. In M. Reite & T. Field (Eds.), *Psychobiology of attachment*. New York: Academic Press.

Fisher, R. A. (1929). Tests of significance in harmonic analysis. *Proceedings of the Royal Society, Series A, 125*, 54–59.

Fiske, D. W., & Rice, L. (1955). Intra-individual response variability. *Psychological Bulletin, 52*, 217–250.

Fox, N. A. (1989). Heart-rate variability and behavioral reactivity: Individual differences in autonomic patterning and their reaction to infant and child temperament. In J. S. Reznick (Ed.), *Perspectives on behavioral inhibition*. Chicago: University of Chicago Press.

Glass, L., & Mackey, M. C. (1988). *From clocks to chaos: The rhythms of life*. Princeton, NJ: Princeton University Press.

Goodwin, B. (1970). Biological stability. In C. H. Waddington (Ed.), *Toward a theoretical biology* (Vol. 3). Chicago: Aldine.

Gottman, J. M. (1981a). Detecting cyclicity in social interaction. *Psychological Bulletin, 86*, 335–348.

Gottman, J. M. (1981b). *Time-series analysis: A comprehensive introduction for social sciences*. Cambridge, UK: Cambridge University Press.

Gottman, J. M., & Levenson, R. W. (1985). A valid procedure for obtaining self-report of affect in marital interaction. *Journal of Consulting and Clinical Psychology, 53*, 151–160.

Gottman, J. M., & Ringland, J. T. (1981). The analysis of dominance and bidirectionality in social development. *Child Development, 52*, 393–412.

Gottman, J. M., & Roy, A. K. (1990). *Sequential analysis: A guide for behavioral researchers*. New York: Cambridge University Press.

Gregson, R. A. M. (1983). *Time series in psychology*. Hillsdale, NJ: Erlbaum.

Grenander, V., & Rosenblatt, M. (1957). *Statistical analysis of stationary time series*. New York: Wiley.

Hamming, R. W. (1983). *Digital filters* (2nd ed.). Englewood Cliffs, NJ: Prentice-Hall.

Hammond, B. R., Jr., Warner, R. M., & Fuld, K. (1995). Blood pressure and sensitivity to flicker. *Journal of Psychophysiology, 9*, 212–220.

Hayes, D. T., & Cobb, L. (1979). Ultradian rhythms in social interaction. In A. W. Siegman & S. Feldstein (Eds.), *Of speech and time: Temporal speech rhythms in interpersonal contexts*. Hillsdale, NJ: Erlbaum.

Healy, D., & Williams, J. M. G. (1988). Dysrhythmia, dysphoria, and depression: The interaction of learned helplessness and circadian dysrhythmia in the pathogenesis of depression. *Psychological Bulletin, 103*, 163–178.

Hepworth, J. T., & West, S. G. (1988). Lynchings and the economy: A time-series reanalysis of Hovland and Sears [1940]. *Journal of Personality and Social Psychology, 55*, 239–247.

Hlastala, M. P., Wranne, B., & Lenfant, C. (1973). Cyclical variations in FRC and other respiratory variables in resting man. *Journal of Applied Physiology, 34*, 670–676.

Hofer, M. A. (1984). Relationships as regulators: A psychobiological perspective on bereavement. *Psychosomatic Medicine, 46*, 183–197.

Hovland, C. J., & Sears, R. R. (1940). Minor studies in aggression: VI. Correlation of lynchings with economic indices. *Journal of Psychology, 9*, 301–310.

Iberall, A. S., & McCulloch, W. S. (1969). The organizing principle of complex living systems. *Journal of Basic Engineering, 91*, 290–294.

Jaffe, J., & Feldstein, S. (1970). *Rhythms of dialogue.* New York: Academic Press.

Jaffe, J., Stern, A. N., & Peery, J. C. (1973). "Conversational" coupling of gaze behavior in prelinguistic human development. *Journal of Psycholinguistic Resarch, 2*, 321–328.

Jenkins, G. M., & Watts, D. G. (1968). *Spectral analysis and its applications.* San Francisco: Holden-Day.

Kaplan, H. L. (1983). Correlations, contrasts, and components: Fourier methods in a more familiar terminology. *Behavior Research Methods and Instrumentation, 15*, 228–241.

Kenny, D. A., & Judd, C. M. (1996). A general procedure for the estimation of interdependence. *Psychological Bulletin, 119*, 138–148.

Koopmans, L. H. (1995). *The spectral analysis of time series.* San Diego, CA: Academic Press.

Kruse, J. A., & Gottman, J. M. (1982). Time series methodology in the study of sexual hormonal and behavioral cycles. *Archives of Sexual Behavior, 11*, 405–415.

Kushner, H., & Falkner, B. (1981). A harmonic analysis of cardiac response of normotensive and hypertensive adolescents during stress. *Journal of Human Stress, 7*, 21–27.

Larsen, R., & Kasimatis, M. (1990). Individual differences in entrainment of mood to the weekly calendar. *Journal of Personality and Social Psychology, 58*, 164–171.

Lester, B. M., Hoffman, J., & Brazelton, T. B. (1985). The rhythmic structure of mother–infant interaction in term and preterm infants. *Child Development, 56*, 15–27.

Linden, W. (1987). A microanalysis of autonomic activity during human speech. *Psychosomatic Medicine, 49*, 562–578.

McCleary, R., & Hay, R. A., Jr. (1980). *Applied time series analysis for the social sciences.* Beverly Hills, CA: Sage.

McClintock, M. (1971). Menstrual synchrony and suppression. *Nature, 229*, 244–245.

Moore-Ede, M. C., Sulzman, F. M., & Fuller, C. A. (1982). *The clocks that time us*. Cambridge, MA: Harvard University Press.

Murray, J. D. (1993). *Mathematical biology* (2nd, corrected ed.). New York: Springer-Verlag.

Norusis, M. J. (1993). *SPSS for Windows Base System Users Guide Release 6.0.* Chicago: SPSS, Inc.

Ostrom, C. W., Jr. (1978). *Quantitative applications in the social sciences: No. 9. Time series analysis: Regression techniques*. Sage Series: Beverly Hills, CA: Sage.

Pool, R. (1989). Is it healthy to be chaotic? *Science, 243,* 604.

Porges, S. W., Bohrer, R. E., Cheung, M. N., Drasgow, F., McCabe, P. M., & Keren, G. (1980). New time-series statistic for detecting rhythmic co-occurrence in the frequency domain: The weighted coherence and its application to psychophysiological research. *Psychological Bulletin, 88,* 580–587.

Porges, S. W., & Coles, M. G. (1982). Individual differences in respiratory–heart period coupling and heart period responses during two attention-demanding tasks. *Physiological Psychology, 10,* 215–220.

Robertson, S. S., Dierker, L. J., Sorokin, Y., & Rosen, M. G. (1982). Human fetal movement: Spontaneous oscillations near one cycle per minute. *Science, 218*(4579), 1327–1330.

Russell, R. J. (1985). Significance tables for the results of Fast Fourier Transforms. *British Journal of Mathematical and Statistical Psychology, 38,* 116–119.

SPSS, Inc. (1993) *SPSS for Windows TRENDS, Release 6.0.* Chicago: Author.

Tabachnick, B. G., & Fidell, L. S. (1996). *Using multivariate statistics* (3rd ed.). New York: HarperCollins.

Tronick. E. Z. (1986). Interactive mismatch and repair: Challenges to the coping infant. *Zero to Three: Bulletin of the National Center for Clinical Infant Programs, 3,* 1–6.

Tronick, E. Z., Als, H., & Brazelton, T. B. (1980). Monadic phases: A structural descriptive analysis of infant–mother face-to-face interaction. *Merrill-Palmer Quarterly, 26,* 3–24.

VanLear, C. A. (1991). Testing a cyclical model of communicative openness in relationship development: Two longitudinal studies. *Communication Monographs, 58,* 337–361.

Wade, M. G., Ellis, M. J., & Bohrer, R. E. (1973). Biorhythms in the activity of children during free play. *Journal of the Experimental Analysis of Behavior, 20,* 155–162.

Warner, R. M. (1988). Rhythm in social interaction. In J. E. McGrath (Ed.), *The social psychology of time: New perspectives*. Beverly Hills, CA: Sage.

Warner, R. M. (1991). Incorporating time. In B. Montgomery & S. Duck (Eds.), *Studying interpersonal interaction*. New York: Guilford Press.

Warner, R. M. (1992a). Cyclicity of vocal activity increases during conversation: Support for a nonlinear systems model of dyadic social interaction. *Behavioral Science, 37,* 128–138.

Warner, R. M. (1992b). Sequential analysis of social interaction: Assessing internal versus social determinants of behavior. *Journal of Personality and Social Psychology, 63,* 51–60.

Warner, R. M. (1992c). Speaker, partner and observer evaluations of affect during social interaction as a function of interaction tempo. *Journal of Language and Social Psychology, 11*, 1–14.

Warner, R. M. (1996). Coordinated cycles in behavior and physiology during face-to-face social interactions. In J. H. Watt & C. A. VanLear (Eds.), *Dynamic patterns in communication processes*. Thousand Oaks, CA: Sage.

Warner, R. M., Malloy, D., Schneider, K., Knoth, R., & Wilder, B. (1987). Rhythmic organization of social interaction and observer ratings of affect and involvement. *Journal of Nonverbal Behavior, 11*, 57–74.

Warner, R. M., & Stevens, A. (1991, October 10). *Cyclic variations in blood pressure during baseline and conversation*. Paper presented at the meeting of the Society for Psychphysiological Research, Chicago.

Warner, R. M., Waggener, T. B., & Kronauer, R. E. (1983). Synchronized cycles in ventilation and vocal activity during spontaneous conversational speech. *Journal of Applied Physiology: Respiration, Environmental and Exercise Physiology, 54*, 1324–1334.

Webb, W. (Ed.). (1982). *Biological rhythms, sleep, and performance*. Chichester, UK: Wiley.

Weber, E. J., Molenaar, P. C., & Van der Molen, M. W. (1992). A nonstationarity test for the spectral analysis of physiological time series data with an application to respiratory sinus arrhythmia. *Psychophysiology, 29*, 55–65.

West, S. G., & Hepworth, J. T. (1991). Statistical issues in the study of temporal data: Daily experiences. [Special issue: Personality and daily experience.] *Journal of Personality, 59*, 609–662.

Winfree, A. T. (1975). Unclocklike behavior of a biological clock. *Nature, 253*, 315–319.

Winfree, A. T. (1983). Sudden cardiac death: A problem in topology. *Scientific American, 248*(5), 144–161.

Index